IFMBE Proceedings

Volume 61

Series editor

Ratko Magjarevic

Deputy Editors

Fatimah Ibrahim
Igor Lacković
Piotr Ładyżyński
Emilio Sacristan Rock

The International Federation for Medical and Biological Engineering, IFMBE, is a federation of national and transnational organizations representing internationally the interests of medical and biological engineering and sciences. The IFMBE is a non-profit organization fostering the creation, dissemination and application of medical and biological engineering knowledge and the management of technology for improved health and quality of life. Its activities include participation in the formulation of public policy and the dissemination of information through publications and forums. Within the field of medical, clinical, and biological engineering, IFMBE's aims are to encourage research and the application of knowledge, and to disseminate information and promote collaboration. The objectives of the IFMBE are scientific, technological, literary, and educational.

The IFMBE is a WHO accredited NGO covering the full range of biomedical and clinical engineering, healthcare, healthcare technology and management. It is representing through its 60 member societies some 120.000 professionals involved in the various issues of improved health and health care delivery.

IFMBE Officers
President: James Goh, Vice-President: Shankhar M. Krishnan
Past President: Ratko Magjarevic
Treasurer: Marc Nyssen, Secretary-General: Kang Ping LIN
http://www.ifmbe.org

More information about this series at http://www.springer.com/series/7403

James Goh · Chwee Teck Lim
Hwa Liang Leo (Eds.)

The 16th International Conference on Biomedical Engineering

ICBME 2016, 7th to 10th December 2016, Singapore

Springer

Editors
James Goh
Department of Biomedical Engineering
National University of Singapore
Singapore

Chwee Teck Lim
Department of Biomedical Engineering
National University of Singapore
Singapore

Hwa Liang Leo
Department of Biomedical Engineering
National University of Singapore
Singapore

ISSN 1680-0737 ISSN 1433-9277 (electronic)
IFMBE Proceedings
ISBN 978-981-10-4219-5 ISBN 978-981-10-4220-1 (eBook)
DOI 10.1007/978-981-10-4220-1

Library of Congress Control Number: 2017935840

Printed on acid-free paper

This Springer imprint is published by Springer Nature
The registered company is Springer Nature Singapore Pte Ltd.
The registered company address is: 152 Beach Road, #21-01/04 Gateway East, Singapore 189721, Singapore

Preface

The International Conference on Biomedical Engineering is a series of international conference in biomedical engineering held in Singapore and is jointly organised by the Faculty of Engineering of the National University of Singapore (NUS) and the Biomedical Engineering Society (Singapore) (BES). The objective of ICBME 2016, held from 7 – 10 December, was to provide local and international participants an invaluable opportunity to stay current on the latest scientific developments and emerging challenges of the biomedical engineering field.

Biomedical engineering is applied in most aspects of our healthcare ecosystem. From electronic health records to diagnostic tools to therapeutic, rehabilitative and regenerative treatments, the work of biomedical engineers is evident. Biomedical engineers work at the intersection of engineering, life sciences and healthcare. The engineers would use principles from applied science including mechanical, electrical, chemical and computer engineering together with physical sciences including physics, chemistry and mathematics to apply them to biology and medicine. Applying such concepts to the human body is very much the same concepts that go into building and programming a machine. The goal is to better understand, replace or fix a target system to ultimately improve the quality of healthcare.

ICBME 2016 received some 350 abstracts from 25 countries and the programme featured more than 120 speakers who spoke on new research and developments in six themes including BioImaging and BioSignals, Bio-Micro/Nanotechnologies, BioRobotics and Medical Devices, Biomaterials and Regenerative Medicine, BioMechanics and Mechanobiology, Engineering/Synthetic Biology across the four-day programme comprising 60 sessions.

Special sessions included the topics of Advances in Microfluidics and Nanofluidics, Bioelectronic Devices, Bioengineering the Heart, Bioimaging, Biomedical Engineering/Clinical Engineering, Biomedical Nanotechnology, Cardiovascular Flows, Engineering Biology/ Synthetic Biology, Flexible and Wearable Technologies, Health Informatics, IFMBE Health Technology Assessment, Integrated Nano-biomechanics: Biological Flow, Mechanobiology, Rehabilitation Robotics, Stem Cells and Organs-on-Chips, Surgical Robotics and Navigation.

The programme was also headlined by seven plenary lectures which include "Cell Therapy of Diabetes Mellitus and its Complications: How Far are we?" by Bernat Soria Escoms, "Bio-integrated and Bio-inspired Stretchable Electronics" by Yonggang Huang, "Recent Advances on Nature Inspired Tissue Engineering Approaches for the Regeneration of Different Tissues" by Rui L. Reis, "Innovative Healthcare is in the Palm of your Hand" by Luke Lee, " Droplet Microfluidics for Single Cell Studies" by David A. Weitz, "Making MRI Safe for Implanted Devices" by John M. Pauly, and "Need Driven Innovation in Medical Technology: From I to I" by Lawrence Ho.

ICBME has attracted a greater number of participants on every occasion that it has met and we are proud to have received the continued endorsement of International Federation for Medical and Biological Engineering (IFMBE) and for the first time, the endorsement of the Agency for Science, Technology and Research (A*STAR), as well as those of many local and regional societies. Their support is a testament to the quality of the conference and we are heartened to have the participation from an impressive roster of highly respected and internationally renowned speakers to lead the programme. The faculty represented the region's best of biomedical engineering and we hope it has provided an excellent opportunity for our fellow colleagues to contribute research and to keep abreast of the exciting developments in our field.

February 2017

Prof. James GOH
Chairman
Prof. Chwee Teck LIM
Scientific Programme Chair
Assoc. Prof. Hwa Liang LEO
Scientific Programme Co-Chair
16th ICBME Organising Committee

The original version of the book was revised.
Incorrect book title has been updated.
The erratum to the book is available at
10.1007/978-981-10-4220-1_25

Acknowledgements

Organised by

National University of Singapore

Biomedical Engineering Society (Singapore)

Supporting Partners

Agency for Science, Technology and Research

IFMBE

Supporting Organizations

Biomedical Engineering Society

Consortium for Medical Device Technologies

The Korean Society of Medical & Biological Engineering

The Institution of Engineers Singapore

Society for Biomaterials

Springer

Sponsored by

AIP Publishing

Supporting Media

Bioengineering

Organization

Organising Committee

Chairperson	James Goh
Co-chair & Treasurer	Siew Lok Toh
Scientific Programme Chair	Chwee Teck Lim
Scientific Programme Co-chair	Hwa Liang Leo

International Advisory Panel

Kai-Nan An	Mayo Clinic
Dominique Barthes-Biesel	Université de Technologie de Compiegne
Shu Chien	University of California San Diego
Yubo Fan	Beihang University
Jing Fang	Peking University
Benny Geiger	Weizmann Institute of Science
Peter Hunter	University of Auckland
Zhonglai Jiang	Shanghai Jiao Tong University
Roger Kamm	Massachusetts Institute of Technology
Andrew F. Laine	Columbia University
Mark Pearcy	Queensland University of Technology
Fong-Chin Su	National Cheng Kung University
Geert Schmid-Schobein	University of California San Diego
David F. Williams	Wake Forest Institute of Regenerative Medicine
Takami Yamaguchi	Tohoku University

International Scientific Committee

Taiji Adachi	Kyoto University
Gang Bao	Rice University
Rashid Bashir	University of Illinois Urbana-Champaign
Markus J. Buehler	Massachusetts Institute of Technology
Anthony M.J. Bull	Imperial College London
Hsueh-Chia Chang	University of Notre Dame
Cheng-Kung Cheng	National Yang Ming University
Mian Long	Institute of Mechanics, Chinese Academy of Sciences
Nigel Lovell	University of New South Wales
Ratko Magjarević	University of Zagreb
Takeo Matsumoto	Nagoya University
Josep Samitier	Institute for Bioengineering of Catalonia
Karol Miller	University of Western Australia
Arthur Mak	Chinese University of Hong Kong
Dennis Discher	University of Pennsylvania
David Elad	Tel Aviv University
C. Ross Ethier	Georgia Institute of Technology
Shih-Kang Fan	National Taiwan University
Huajian Gao	Brown University

Amit Gefen	Tel Aviv University
Guy Genin	Washington University in St. Louis
J. Justin Gooding	University of New South Wales
Farshid Guilak	Washington University in St. Louis
X. Ed Guo	Columbia University
Walter Herzog	University of Calgary
Gerhard Holzapfel	Graz University of Technology
Dean Ho	University of California Los Angeles
I-Ming Hsing	Hong Kong University of Science and Technology
Fatimah Binti Ibrahim	University of Malaya
Takuji Ishikawa	Tohoku University
Noo Li Jeon	Seoul National University
Michael Khoo	University of Southern California
Eng Hin Lee	National University of Singapore
Sang-Hoon Lee	Korea University
Marcus Pandy	University of Melbourne
Heinz Redl	Ludwig Boltzmann Institute of Experimental and Clinical Traumatology
Robert O. Ritchie	University of California Berkeley
John A. Rogers	University of Illinois Urbana-Champaign
Masaaki Sato	Tohoku University
Masahiro Sokabe	Nagoya University
Joachim P. Spatz	University of Heidelberg
Masao Tanaka	Osaka University
Ichiro Sakuma	University of Tokyo
Sehyun Shin	Korea University
Jackrit Suthakom	Mahidol University
Nitish Thakor	National University of Singapore
David W. Townsend	A*STAR-NUS Clinical Imaging Research Centre
Herbert Voigt	Boston University
Shigeo Wada	Osaka University
Jaw-Lin Wang	National Taiwan University
David A. Weitz	Harvard University
Ed X. Wu	Hong Kong University
Chaoyong Yang	Xiamen University
Guang-Zhong Yang	Imperial College London
Kam Leong	Columbia University
Feng-Huei Lin	National Taiwan University
Kang-Ping Lin	Chung Yuan Christian University
Hao Liu	Chiba University
Ajit Yoganathan	Georgia Institute of Technology
Sun Yu	University of Toronto
Haishan Zeng	University of British Columbia

Local Scientific Committee

Scientific Programme Chair

Chwee Teck Lim	National University of Singapore

Scientific Programme Co-chair

Hwa Liang Leo	National University of Singapore

Members

Themes and Topics

A: Bioimaging and Biosignals

1. Bioinformatics & Digital Medicine
2. Biomedical Imaging
3. Biomedical instrumentation
4. Biosignal Processing
5. Clinical Engineering
6. Neuroengineering
7. Telemedicine & Healthcare

B: Bio-micro/Nanotechnologies

1. Biosensors, BioMEMs & Lab-on-Chip
2. Micro/Nano Biomedical Devices & Systems
3. Nanobiotechnology

C: BioRobotics and Medical Devices

1. Computer-Integrated & Computer-Assisted Surgery
2. Ergonomics & Human Factors
3. Medical Robotics
4. Rehabilitation Engineering & Assistive Technology
5. Wearable Devices

D: Biomaterials and Regenerative Medicine

1. Artificial Organs
2. Biomaterials
3. Controlled Drug Delivery
4. Pharmaceutical Sciences & Engineering
5. Regenerative Medicine & Tissue Engineering
6. 3D Bioprinting

E: BioMechanics and Mechanobiology

1. Cardiovascular Bioengineering
2. Cell & Molecular Bioengineering
3. Computational Bioengineering
4. Organ and Tissue Mechanics
5. Orthopaedic Biomechanics
6. Mechanobiology
7. Sports Biomechanics & Human Performance

F: Engineering/Synthetic Biology

1. Cell-free synthetic biology and engineering synthetic ecosystems
2. Microbial production of drugs
3. Artificial cells
4. Synthetic viruses
5. Genetic engineering
6. Artificial photosynthesis

Contents

Development of an intelligent pillow with multiple flexible actuators. . . . 1
Akisue Kuramoto, Wataru Inoue, Yasuhito Otake, Hitoshi Kimura, Norio Inou, Sacha Krivokuca, Tomu Ichikawa, Hiroyuki Ono, and Naoto Sekiyama

Objective Evaluation of Skin Texture Condition by Image Analysis 7
Toshiyuki Tanaka and Haruna Suzuta

Effects of Photofrin-mediated photodynamic treatment on sensitivity to cisplatin in HeLa cells and the resistant subline. . . . 13
Yuichi Miyamoto, Yuri Yamaguchi, and Hatsuki Tajima

Novel spectral approach for pulse waves contour analysis 16
A.A. Fedotov, A.S. Akulova, S.A. Akulov, and M.A. Reshetnikova

A Comprehensive Analysis on Breast Cancer Classification with Radial Basis Function and Gaussian Mixture Model 21
Harikumar Rajaguru and Sunil Kumar Prabhakar

Heart Sounds Features Usage for Classification of Ventricular Septal Defect Size in Children. . . . 28
Kamran Hassani, Kamal Jafarian, and D. John Doyle

Multiclass Classification of APG Signals using ELM for CVD Risk Identification: A Real-Time Application 32
B. Niranjana Krupa, Kunal Bharathi, Manjunath Gaonkar, Sai Karun, Suhan Nath, and M.A.M. Ali

Development of an EMG-based Human-Machine Interface on Open-source Linux Platform for Evaluating the Motor Skill Acquisition Process. . . . 38
Guanghao Sun and Wenwei Yu

Measuring Neural, Physiological and Behavioral Effects of Frustration 43
Babu Ram Naidu Ramachandran, Sandra Alejandra Romero Pinto, Jannis Born, Stefan Winkler, and Rama Ratnam

Estimating Correlation between Arterial Blood Pressure and Photoplethysmograph 47
Theertha Abhay, Kayalvizhi N., and Rolant Gini J.

Flow visualization for Nasal Cavity Flow in Aerosol Exhalation Through Nose Treatment. . . . 53
Takahisa Yamamoto, Yoshiki Kobayashi, Shunpei Shikano, Masahiro Takeyama, Mikiya Asako, and Koichi Tomoda

Multi physiological signs model to enhance accuracy of ECG peaks detection 58
A. Delrieu, M. Hoël, C.T. Phua, and G. Lissorgues

PSO based Density Classifier for Mammograms 62
Sneha Simon, Lavanya R., and Devi Vijayan

Development And Evaluation of WEEG: A Wearable 8-Channel System for Brain Computer Interfaces 67
Tri Thong Vo

Performance Analysis of GMM Classifier for Classification of Normal and Abnormal Segments in PPG Signals . . . 73
Sunil Kumar Prabhakar and Harikumar Rajaguru

Development of Patient Remote Monitoring System for Epilepsy Classification 80
Sunil Kumar Prabhakar and Harikumar Rajaguru

Discrete Wavelet Transform based statistical features for the Drowsiness detection from EEG 88
Reddy Vamsi, Dabbu Suman, Nikhil CH, and M. Malini

State space modeling of bioimpedance for haematocrit measurement 95
S.A. Akulov, A.A. Fedotov, A.S. Akulova, and M.A. Reshetnikova

A Hybrid Classifier for the Detection of Microaneurysms in Diabetic Retinal Images 99
Aishwarya R., Vasundhara T., and Ramachandran K.I.

Approach To Extract Twin fECG For Different Cardiac Conditions During Prenatal 106
Rolant Gini J., Ramachandran K.I., and Ceerthibala U.K.

Reception Performance Enhancement of Capacitive Micromachined Ultrasonic Transducers Via Modified Membrane Structures 111
Xiao Jiang, Yuanyu Yu, Jiujiang Wang, Xin Liu, U. Kin Che, Sio Hang Pun, Mang I. Vai, and Peng Un Mak

Design of a Silent Speech Interface using Facial Gesture Recognition and Electromyography 117
Aishwarya Nair, Niranjana Shashikumar, S. Vidhya, and S.K. Kirthika

Performance Comparison of Oral Cancer Classification with Gaussian Mixture Measures and Multi Layer Perceptron 123
Harikumar Rajaguru and Sunil Kumar Prabhakar

Measurement of Pulse Transit Time (PTT) Using Photoplethysmography 130
Lokharan M., Lokesh Kumar K.C., Harish Kumar V., Kayalvizhi N., and Aryalekshmi R.

Erratum to: The 16th International Conference on Biomedical Engineering E1
James Goh, Chwee Teck Lim, and Hwa Liang Leo

Author Index 135

Development of an intelligent pillow with multiple flexible actuators

Akisue Kuramoto[*1], Wataru Inoue[1], Yasuhito Otake[1], Hitoshi Kimura[1], Norio Inou[1], Sacha Krivokuca[2], Tomu Ichikawa[3], Hiroyuki Ono[3], Naoto Sekiyama[3]

[1]Department of Mechanical and Control Engineering, Tokyo Institute of Technology, Tokyo, Japan
[*1]kuramoto.a.aa@m.titech.ac.jp
[2]Polytech Paris UPMC, School of Engineering, Paris, France
[3]Fujibedkogyo Co., Ltd., Tokyo, Japan

Abstract— This study aims to develop an intelligent bedding that automatically provides a comfortable sleep condition for each person. This paper proposes a pillow-type device which actively changes its shape and stiffness. The device is composed of multiple flexible actuators which are driven by air pressure. Each actuator has two types of sensors: an internal pressure sensor and five contact pressure sensors. These sensors are used for monitoring the support condition of user's head and neck. We developed a pillow system capable of controlling the internal pressure and contact pressure of each actuator in order to realize a comfortable condition for sleep. The pillow system successfully controlled the center of contact pressure to the target position. To elucidate comfortable and uncomfortable support conditions, pressure distributions on the head and neck were measured by the system. The result indicates that accurate control of the center of contact pressure is important for comfortableness.

Keywords—sleep; comfortable; intelligent bedding system; flexible actuator; feedback control; pressure; posture

I. INTRODUCTION

Comfortable bedding, especially pillow, is important for health and quality of life. There is a lot of possible factors to provide comfortable pillows. Many researchers tried to evaluate comfortableness of pillows with those factors such as stiffness [1][2], shape [3][4][5][6][7] and thermal properties [8]. Some researchers reported development of an active air mattresses for controlling thermal condition of bedding [9] or preventing bedsore of a patient [10] [11] by pressurizing or depressurizing air bags with predetermined routine. However, those mattresses do not have an active function to transform their shape to provide an appropriate posture for each user. In addition, there are a few discussions of such beddings for healthy people. We expect the contact pressure acting on the head and neck area and the sleep posture are important factors for comfortableness of a pillow because they are related to biomechanical conditions of users.

This study aims to develop an intelligent bedding system that automatically provides a comfortable position and pressure distribution. This paper discusses about a new pillow device as a part of the intelligent bedding system. Although there is a report about a pillow device [12], it focuses to maintain breathing easy during sleep and the viewpoint is different from our purpose. This paper describes the pillow device and comfortable support condition.

II. THE INTELLIGENT PILLOW DEVICE

A. Overview

Figure 1 shows the intelligent pillow system. This system is composed of a transformable pillow, control unit, air valves and air compressor. The details of the parts are described below.

B. Transformable pillow

Figure 2 shows a transformable pillow. The pillow part is composed of eight flexible actuators; three for the upper layer (Fig. 2(a)) and five for the lower layer (Fig. 2(b)). Each actuator is made of a soft rubber tube, which is reinforced with a fibrous cover sheet. The actuator can lift about 600 N weight, representing a human body, without air leak. It has two types of sensors: an internal pressure sensor and five contact pressure sensors. The internal pressure sensor (MIS-2500-015G, Metrodyne Microsystem Corp.) is connected to fluid path of each actuator. Contact pressure sensors (FSR406, Interlink Electronics) are put on a surface of actuator. These sensors are used to monitor pressurized condition and contact condition of the actuator. Each actuator also has two valves for controlling internal pressure. As detailed in the following section, the control unit gives to the valves a command to open or close based on contact pressure distribution and internal pressure value.

J. Goh et al. (eds.), *The 16th International Conference on Biomedical Engineering*,
IFMBE Proceedings 61,
DOI: 10.1007/978-981-10-4220-1_1

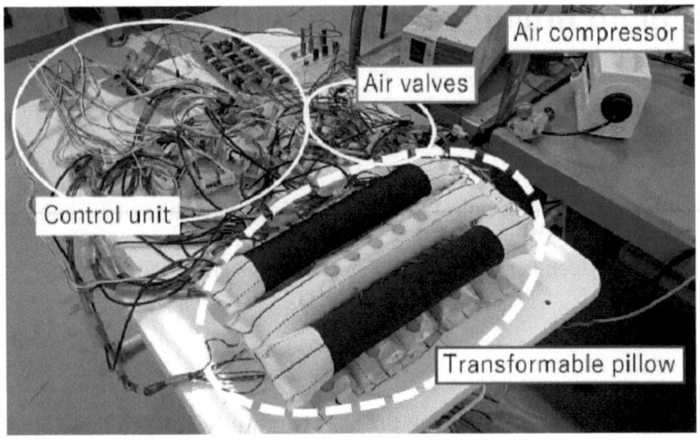

Fig. 1. The intelligent pillow device

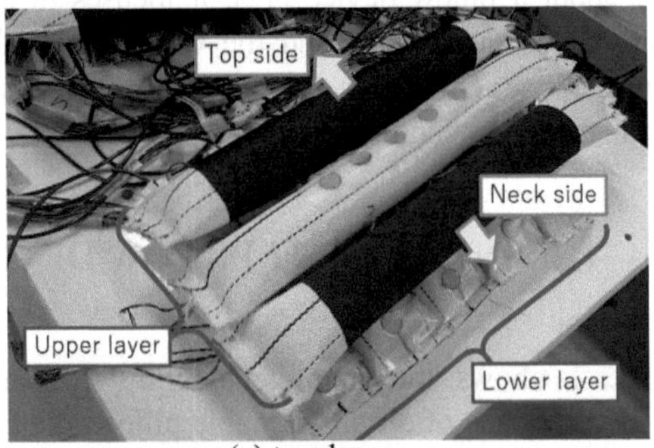

(a) two layers

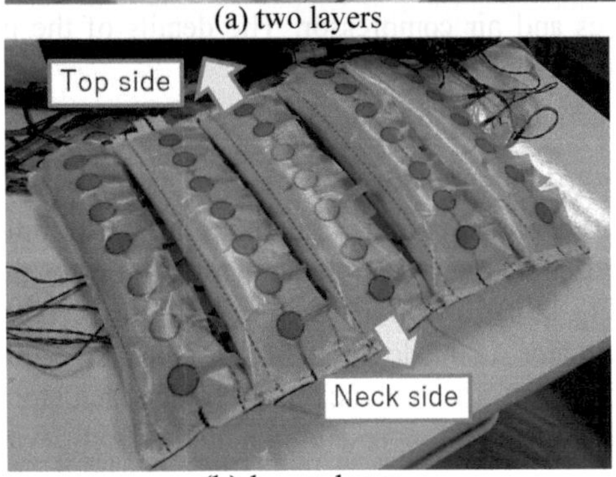

(b) lower layer

Fig. 2. Transformable pillow

The transformable pillow has two actuator layers. The upper three actuators are aligned parallel to the transverse plane of a human body. This layer controls head's pitch angle (neck extension and flexion) as shown in Fig. 3. Neck is extended when a neck side actuator is pressurized. By contrast, neck is flexed when a top side actuator is pressurized. The upper layer is also expected to realize comfortable touch condition.

The lower layer has five actuators, which are aligned parallel to the body trunk. As shown in Fig. 4, the lower layer has two roles; controlling head's yaw angle (neck rotation, Fig. 4(a)) and controlling the whole height of pillow device (Fig. 4(b)). Head turns left when a right side actuator is pressurized. By contrast, head turns right when a left side actuator is pressurized. The height of pillow device can be adjusted by adequately changing the internal pressure of lower layer actuators. With this height change, the neck angle relative to trunk and the head angle relative to neck may be affected lightly. However, these angle changes can be cancelled with controlling upper layer actuators. Therefore, coordinated motion of multiple actuators is necessary for controlling head pitch (neck flexion and extension) and yaw (neck rotation).

C. Control algorithm

Figure 5 shows a fluid path of this pillow device. An air compressor supplies pressurized air to each module of a flexible actuator. Each module has two valves to control internal pressure of an actuator. The valves receive commands to pressurize or depressurize actuator from the control unit.

Figure 6 shows an example of flowchart. This flowchart regulates the current center of mass (CCOM) of head and neck with the target center of mass (TCOM). The CCOM can be calculated from the following equation,

$$\boldsymbol{CCOM}= \frac{\sum_i (p_i l_{rc})}{\sum_i p_i}, \qquad (1)$$

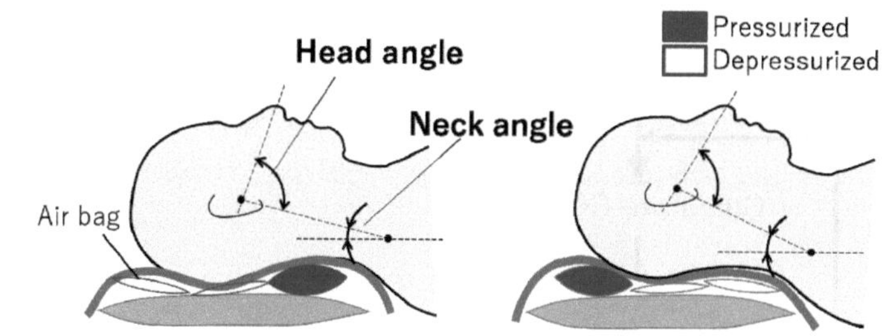

Fig. 3. Role of upper layer actuators

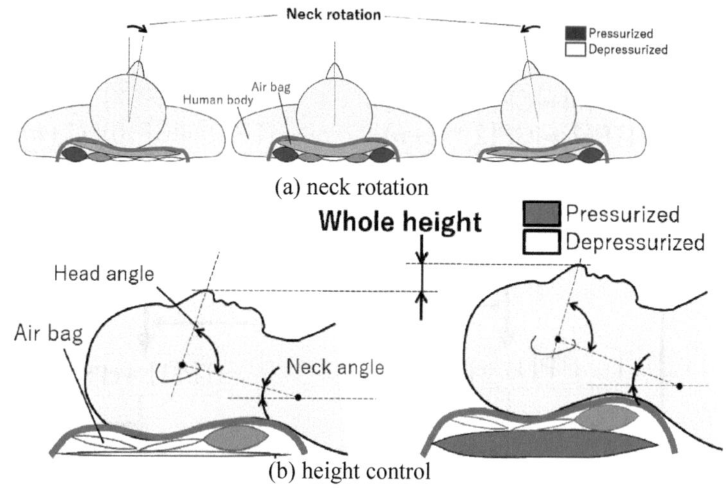

Fig. 4. Two roles of lower layer actuators

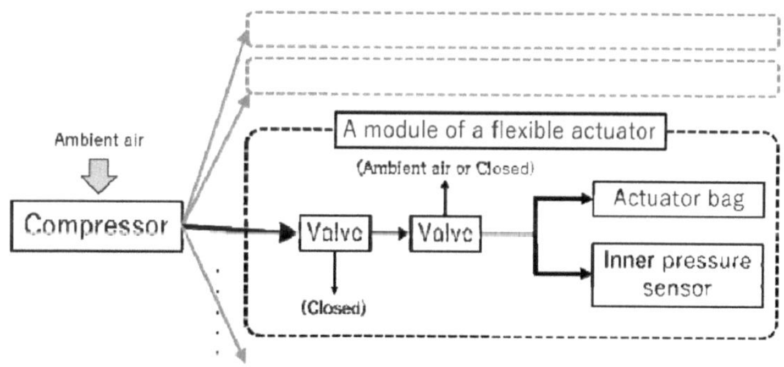

Fig. 5. Schema of fluid path

where, i is a number of contact pressure sensor, p_i is contact pressure value of the ith sensor, and $\boldsymbol{l}_i$ is location of the ith sensor. The distance between CCOM and TCOM is regarded as an error in the control algorithm.

The control system also monitors internal current pressure value (ICP) of each actuator. In the control algorithm, the internal target pressure value (ITP) of next step is set based on both the distance error and the ICP value.

We performed a preliminary test with setting the TCOM manually. The control algorithm works well to manage the distance between CCOM and TCOM less than 10 mm.

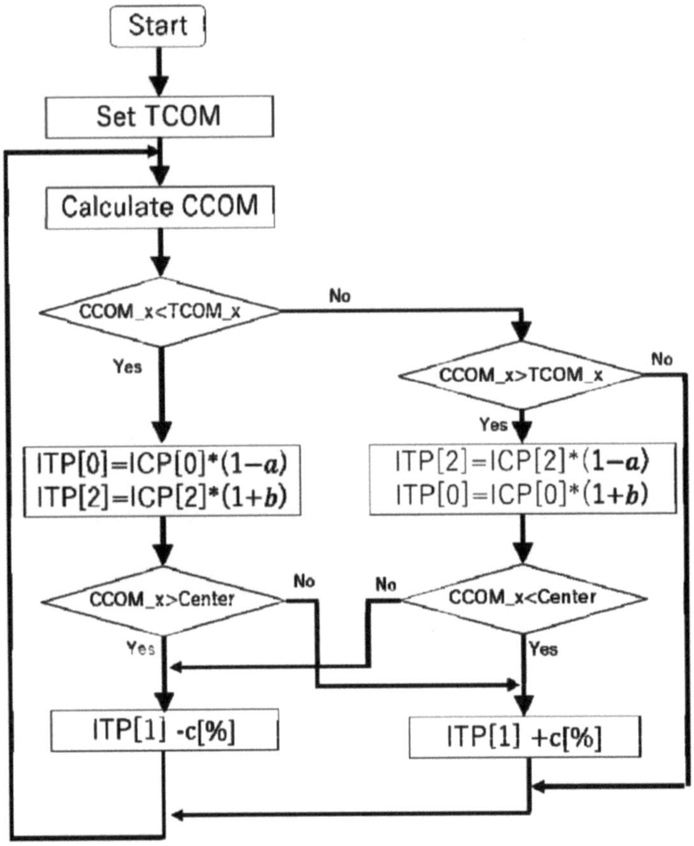

Fig. 6. Example of control flowchart for controlling center of mass of head and neck area

III. EXPERIMENT OF TRANSFORMABLE PILLOW

To drive this intelligent pillow device with the control algorithm described in the previous section, an appropriate setting of the TCOM as reference value of comfortableness is indispensable. However, there are few reports about support condition of head and neck with a pillow for comfortableness. In addition, unfavorable pressure data are also necessary to avoid uncomfortable conditions. As a pilot study, manual operation is used to find comfortable and uncomfortable conditions. As shown in Fig. 7, remote controller for manual operation is prepared for this test. Description of button number and function is shown in Table1. We assumed that a human prefers a symmetrical posture in supine position. To simplify an operation of actuators in this test, the lower layer was used only for controlling height of the device. All actuators of the lower layer are driven in a similar way.

We asked several subjects to manipulate the pillow device as they feel comfortable. When comfortable conditions are determined, contact pressure distribution and internal pressure value of actuators are recorded. Then we searched conditions that subjects feel uncomfortable by changing internal pressure of actuator. Measurement of contact pressure distribution is continued during the test. When a subject feels more discomfort than the temporary worst, the pressure distribution is recorded. At the same time, the actual internal pressure of actuators is also recorded.

The series of Figure 8 show pressure distribution images of a subject's head and neck area measured by the X SENSOR (PX100: 40.64.02, XSENSOR Technology Corp.). The whole pressure distribution images of the head and neck area in the worst setting (uncomfortable) and in the best setting (comfortable) are shown in Fig. 8(a) and (c). Figure 8(b) shows the pressure distribution when the subject uses his comfortable pillow which is commercially available. When the subject feels comfortable (Fig. 8(a) and 8(b)), neck weakly touches the pillow and the head is supported with a wide area. In contrast, when the subject feels uncomfortable (Fig. 8(c)), neck strongly touches the pillow and the pressure peak is on the neck area. This result shows that appropriate support condition is necessary to realize automatic comfortable pillow.

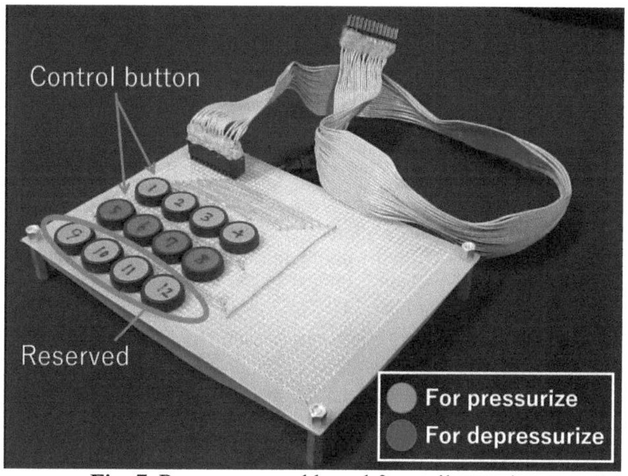

Fig. 7. Remote control board for a pilot study

TABLE I. THE NUMBER AND FUNCTION OF CONTROL BUTTONS

No.	Target actuator	Function
1	Upper, Top-side	Pressurize
2	Upper, Middle	
3	Upper, Neck-side	
4	Lower actuators	
5	Upper, Top-side	Depressurize
6	Upper, Middle	
7	Upper, Neck-side	
8	Lower actuators	
9~12	None	None

When subjects feel comfortable, the CCOM, which is calculated from eq. (1), is located in near the neck area. On the other hand, when subjects feel uncomfortable, the CCOM is located inside the neck area. Although the difference of CCOM locations relative to the body is small, there is a large difference of feeling. This result indicates the importance of controlling the CCOM accurately for comfortableness.

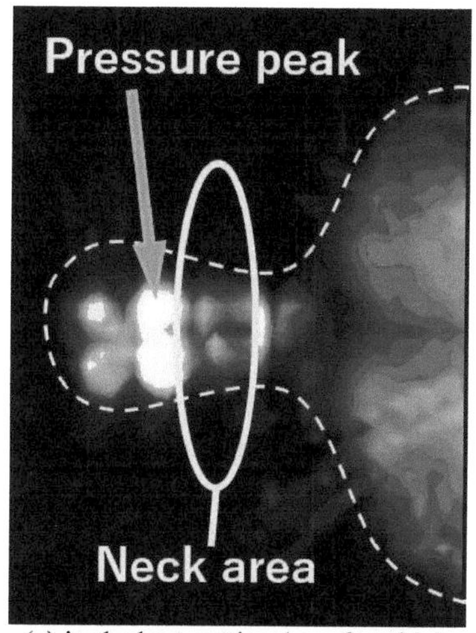

(a) in the best setting (comfortable)

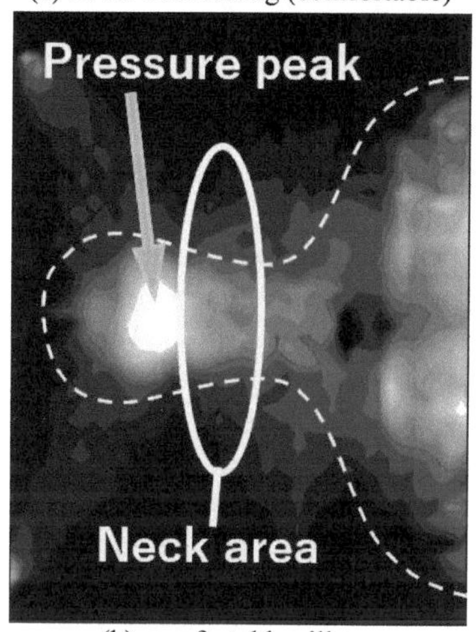

(b) comfortable pillow

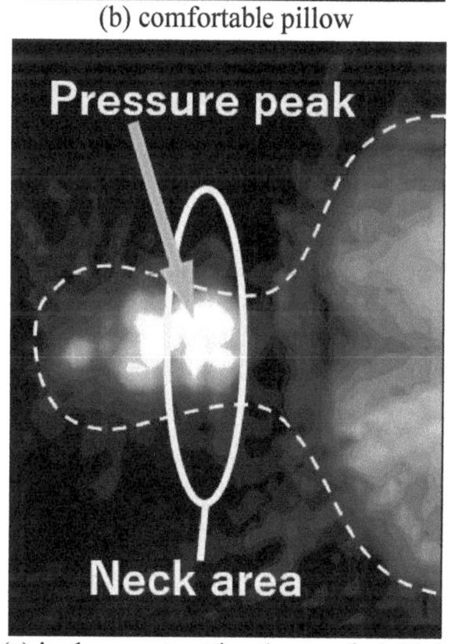

(c) in the worst setting (uncomfortable)

Fig. 8. Pressure distribution of head and neck area (White colored areas are pressurized.)

During our tests, we confirmed that the maximum pressure of neck-side bag should be less than 5 kPa. If it is higher than 5 kPa, the feeling of contact in the neck area is too hard. The user seems to be the most comfortable when the COM is near the neck while having at the same time not too high pressure on the neck-side bag. This information will help to improve the control algorithm as to set the TCOM automatically.

The proposed device is also applicable when customers choose an ordinary pillow. The number of pillows trial to find a good one can be decreased if customers find a comfortable pillow height with this device.

IV. CONCLUSION

This study proposed the intelligent pillow system as a part of an intelligent bedding system. The system is composed of a transformable pillow, control unit, air valves and air compressor. We developed the control algorithm of the pillow device. In head and neck area, the distance between current center of mass (CCOM) and the target center of mass is controlled with less than 10 mm. With several subjects, support conditions for comfortableness and uncomfortableness are also examined. The result indicated that the location of the CCOM relative to human body is important for the comfortableness. Pressurized condition of actuators is also important for a feeling of contact in neck area. Considering a hysteresis loop of actuators should be necessary for a more accurate control of CCOM. Increasing the number of subjects is also necessary for developing the evaluation method of body support condition.

REFERENCES

[1] F.M. Kovacs, V. Abraira, A. Peña, J.G. Martín-Rodríguez, M. Sánchez-Vera, E. Ferrer, D. Ruano, P. Guillén, M. Gestoso, A. Muriel, J. Zamora, M.T. G. del Real, and N. Mufraggi, "Effect of firmness of mattress on chronic non-specific low-back pain: randomised, double-blind, controlled, multicentre trial," Lancet, Vol.362(9396), 2003, pp.1599-1604.

[2] M.B. Scharf, R. Stover, M.McDannold, H. Kaye, and D.V. Berkowitz, "Comparative effects of sleep on a standard mattress to an experimental foam surface on sleep architecture and CAP rates, " Sleep, Vol.20(12), 1997, pp.1197-1200.

[3] R.A. Lavin, M.Pappagallo, K.V. Kuhlemeier. "Cervical pain: A comparison of three pillows". Original Research Article Archives of Physical Medicine and Rehabilitation, 78 (2), pp. 193-198.

[4] P. Erfanian, S. Tenzif, R.C. Guerriero. "Assessing effects of a semi-customized experimental cervical pillow on symptomatic adults with chronic neck pain with and without headache," Journal of the Canadian Chiropractic Association, 48 (1) (2004), pp.20-28.

[5] S.F. Liu, Y.L. Lee, J.C. Liang, "Shape design of an optimal comfortable pillow based on the analytic hierarchy process method," Original Research Article Journal of Chiropractic Medicine, 10 (4) (2011), pp. 229-239.

[6] J.C. Wang, R.C. Chan, H.L. Wu, C.J. Lai, "Effect of pillow size preference on extensor digitorum communis muscle strength and electromyographic activity during maximal contraction in healthy individuals: A pilot study," Journal of the Chinese Medical Association, 78 (3) 2015, pp. 182-187.

[7] D. Caia, H.L. Chen. "Ergonomic approach for pillow concept design," Applied Ergonomics, 52 (2016), pp.142-150.

[8] A. Lin, S. Deng, "A study on the thermal comfort in sleeping environments in the subtropics—Measuring the total insulation values for the bedding systems commonly used in the subtropics," Building and Environment, Vol.43(5), 2008, pp.905-916.

[9] K. Okamoto, K. Mizuno, and N. Okudaira, "The Effects of a Newly Designed Air Mattress upon Sleep and Bed Climate," Journal of Physiological Anthropology, vol.16(4), 1997, pp. 161-166.

[10] A.N. Exton-Smith, J. Wedgwood, P.W. Overstall, and G. Wallace, "Use of the 'Air Wave System' to prevent pressure sores in hospital", Lancet, Vol.319(8284), 1982, pp.1288-1290.

[11] B. Devine, "Alternating Pressure Air Mattresses in the Management of Established Pressure Sores," Journal of Tissue Viability, Vol.5(3), 2014, pp.94-98.

[12] T. Harada, A. Sakata, T. Mori, T. Sato, "Sensor pillow system: Monitoring respiration and body movement in sleep," Proceedings. 2000 IEEE/RSJ International Conference on Intelligent Robots and Systems (IROS 2000), Vol.1, 2000, pp.351-356.

[13] D. Maruyama, H. Kimura, M. Koseki and N. Inou, "Driving force and structural strength evaluation of a flexible mechanical system with a hydrostatic skeleton", Journal of Zhejiang University-SCIENCE A, Applied Physics&Engineering,Vol.11(4), 2010, pp.255-262.

[14] M. Kataoka, H. Kimura and N. Inou, "Hermetically-Sealed Flexible Mobile Robot "MOLOOP" for Narrow Terrain Exploration-Improvement of Flexible Bags with Fibrous Material-", The 39th Annual Conference of the IEEE Industrial Electronics Society(IECON 2013), 2013, pp.4154-4159.

Objective Evaluation of Skin Texture Condition by Image Analysis

Toshiyuki Tanaka[1] and Haruna Suzuta[1]

Keio University, Yokohama, 223-8522, Japan,
tanaka@appi.keio.ac.jp,
WWW home page: http://isp.appi.keio.ac.jp

Abstract. This paper presents an analysis method that extracts the features of " skin texture " and "unevenness of skin color " from skin images. We determine the effective features for the evaluation of skin texture using a two-sample t-test on the cheek and underside of the chin. The results show that the total number of cristae cutis as well as the mean and standard deviation of their area are effective features for skin texture analysis. In addition, the maximum value of an a* image as well as the standard deviation and kurtosis value of a b* image are effective features for assessing the unevenness of skin color. The results of tests show that the standard deviation of the area with changing colors is an effective feature for skin texture analysis.

Keywords: skin analysis, skin texture, color unevenness, blotch, skin pore, cheek, underside of the chin

1 Introduction

Conventional skin texture evaluation is mostly performed by skilled beauty technicians, and is performed using visual assessment. In recent years, skin texture can be measured objectively using the optical characteristecs of the skin and newly developed measurement approaches [1-5]. However, even now, most systems require great care and are not suitable for general use. Moreover, it is difficult to extract skin texture, area, and color unevenness and assess the skin quantitatively.

In this paper, we focus on "skin texture" and "color unevenness" of skin images, which characterize the skin condition well. Further, we aim to determine the quantitative features that are available for measuring skin texture. An intensity image was investigated for skin texture analysis and a polarized light image was investigated for color unevenness in [6]. Fine body hair and skin pores affect color unevenness [7, 8]. We use frequency information [9] to extract skin pores. New approaches are proposed for the extraction of color information and color assessment, and effective features are selected using discriminant analysis. Moreover, objective assessment is performed by comparing the skin of the cheek with that of the chin.

2 Skin Texture Experiments

2.1 Photographing Skin

In this study, a microscope (USB Microscope M3, Scalar Corporation) [10] is used for taking skin images (Fig. 1). The resolusion of the images is 1, 280 x 1, 024 pixels. As shown in Fig. 2, the LED of this microscope has a polarized light filter and controls the light reflected from the object. Texture images of the skin surface, which can be concave or convex, are clearly obtained when the polarized filter is used. Blemish information is better obtained when the polarized filter is not used. Both images are required to evaluate the texture of skin.

2.2 Imaging Regions

The condition of the skin degrades as each person ages. However, these conditions are different at different sites. We define the skin on the cheek to be the averageskin condition on the face and the skin under the chin to be the most youthful part. The cheek and chin parts are photographed for each subjects using both the polarized and non-polarized modes of the microscope. Figure 3 shows an example of cheek and chin images of the same person. The chin image has better condition than the cheek image with respect to skin texture. The image format is 32-bit PNG, and the resolution of the image is 458 x 334 pixels.

2.3 Subjects of the Experiments

In this study, all the subjects were graduate school students: 16 males and 20 females. We first preprocessed the image. Because we determined that male facial hair affects the subsequent results, we omitted the images of male faces, and analyzed the skin texture of females only.

2.4 Experiment Conditions

Skin condition is greatly affected by the weather. In the present study, we collected skin images in the autumn. The date collection covered a period of two days, except for the part of the photo.

J. Goh et al. (eds.), *The 16th International Conference on Biomedical Engineering*,
IFMBE Proceedings 61,
DOI: 10.1007/978-981-10-4220-1_2

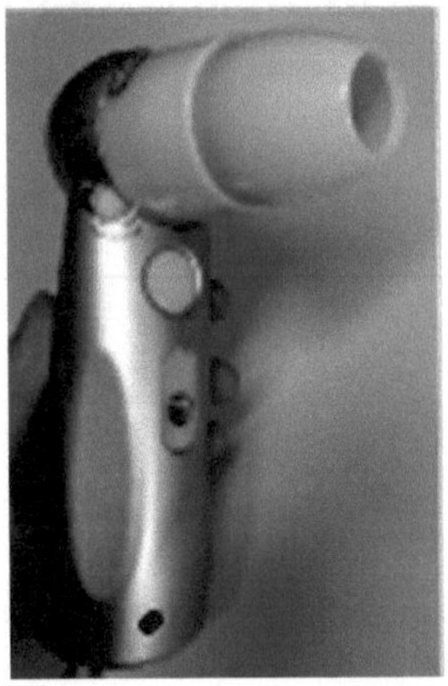

Fig. 1. USB Microscope M3

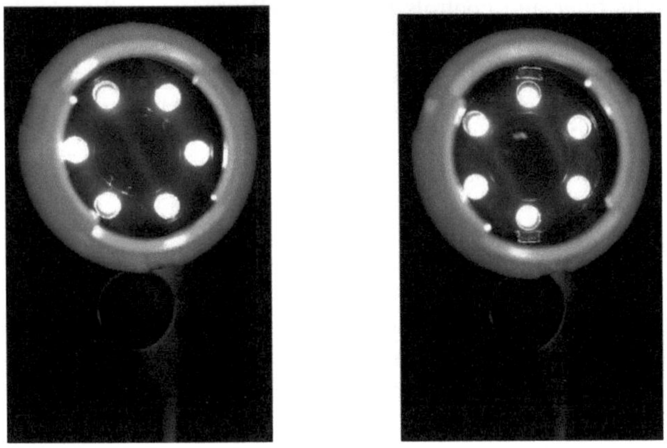

Fig. 2. Mechanism of lighting.

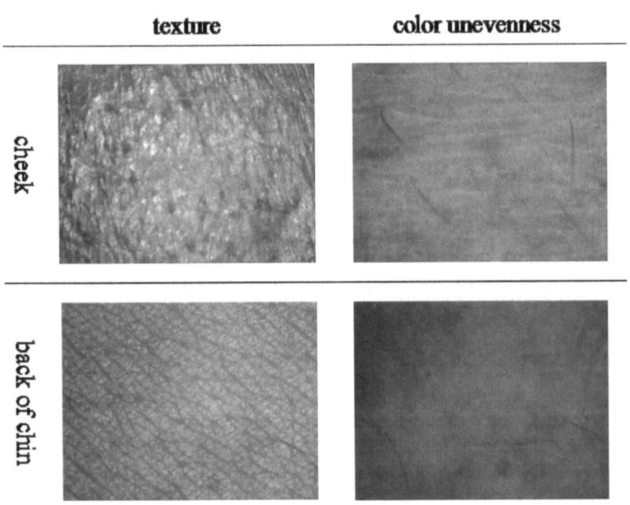

Fig. 3. Example of cheek and chin images

Day 1: 2014 November 13 (10:00-19:00)
Day 2: 2014 November 14 (10:00-19:00)
The average air temperature and precipitation on these days were as follows.
Day 1: Average air temperature 16.97 °C, precipitation 0.0 mm
Day 2: Average air temperature 14.35 °C, precipitation 0.0 mm

3 Proposed Method for Skin Texture Measurement

Figure 4 shows the flowchart of the proposed method. First, texture and color unevenness images are collected by the microscope. The images are preprocessed by filtering and the skin features are quantitatively computed from the obtained images. Discriminant analysis is performed using the computed skin features. Finally, the number, fineness, and proportionality of the skin grit are obtained from the final skin grit images. The color unevenness and spots on the skin are computed from the skin grit images.

3.1 Skin grit

The following four preprocessing steps performed to obtain skin grit images.

(1) Conversion of the image from RGB color space to L*a*b* color space
(2) Adaptive binarization for the L* component image
(3) Component labeling
(4) Removal of the outliers of the feature values

Each of these preprocessing steps are described below.

(1) Conversion of the image from RGB color space to L*a*b* color space

It is difficult to extract the skin grit from an image in RGB color space, because the difference in hue between grooves in the skin and pichu on the skin is small. In this study, we focus on the difference of the intensity between the grooves and pichu on the skin using CIE1976 L*a*b* color space.

(2) Adaptive binarization

Because the light on the head of the camera illuminates the area that is imaged, the obtained images have color unevenness. It is difficult to extract the features using a fixed threshold on such images. Therefore, adaptive binarization is used, in which the threshold changes according to each pixel of the input image.

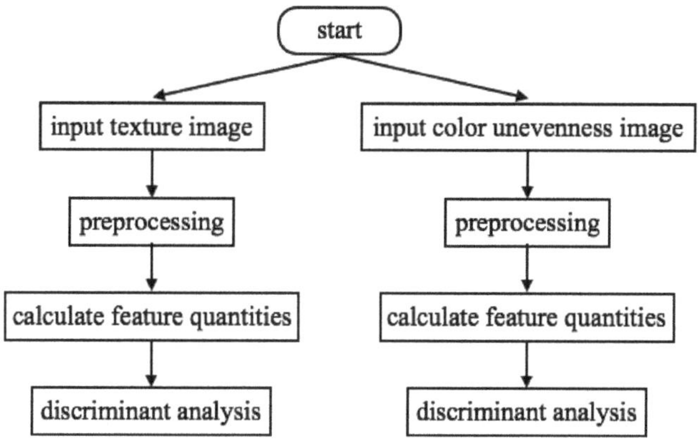

Fig. 4. Flowchart of the proposed method

We denote a Gaussian filter of size of N as M. The adaptive binarization $I_{adaptive}$ of an arbitrary pixel *(x, y)* is computed by

$$I_{\mathbf{adaptive}}(x,y) = M \otimes I(x,y) - 2. \quad (1)$$

In the (1), ⊗ is the filter operation on arbitrary pixel I(x,y). A value of 1 is assigned to the pixel if the intensity is more than the threshold, and a value of 0 is assigned to the pixel otherwise. The subtracted constant reduces the effects of noise and flicker in the background region, specially when the intensity of the near-field region resembles that of the object. In this study, N =15 x 15 pixels and the subtraction constant is set to be 2 empirically.

(3) Component labeling

Because there are many connected components in one image, the components must be numbered for the subsequent processing. Here, the component labeling is performed to number each connected component.

(4) Removal of the outliers of the feature values

The regions with an area of more than 500 pixels and less than 5 pixels are removed from the set of labeled regions. Regions with an area of more than 500 pixel correspond to unsuitably illuminated regions near the edge of the image, and regions with an area of less than 5 pixels correspond to regions that are much smaller than skin grit. The threshold levels are decided empirically using comparison with the original image. Figure 5 shows examples of the obtained skin grit.

In Figure 5, the red lines indicate the connected components after the removal of outlier connected components. The total number of objects (skin grit) as well as the average and standard deviation of the component areas are computedfrom the preprocessed images. These values correspond to the proportionality index of the skin grit.

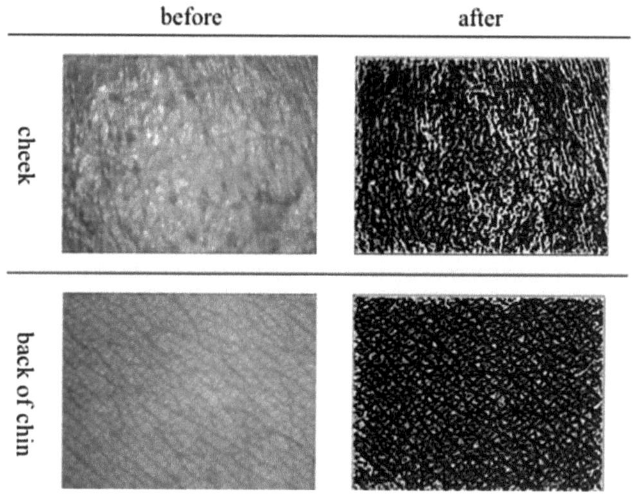

Fig. 5. Example of skin grit extraction

3.2 Color unevenness of the skin

The conversion of RGB images into L*a*b* and the removal of noise (fine facial hair) are performed before the evaluation of the color unevenness. In the following, the noise removal is described.

Especially in chin images, fine facial hair sometimes influences the subsequent results. Therefore, we formulate the process of fine hair removal from the image as noise removal. First, a mask image is constructed for the fine hair, and is superimposed on the original image. The obtained image is interpolated with respect to the fine hair parts, as shown in Fig. 6. In this figure, the red pixel is the reference pixel. The average value is assigned to the reference pixel by raster scanning the whole image and by checking the skin density of the nearest neighbors.

The maximum value, standard deviation, kurtosis, and skewness of the pixel density are computed as the color unevenness features.

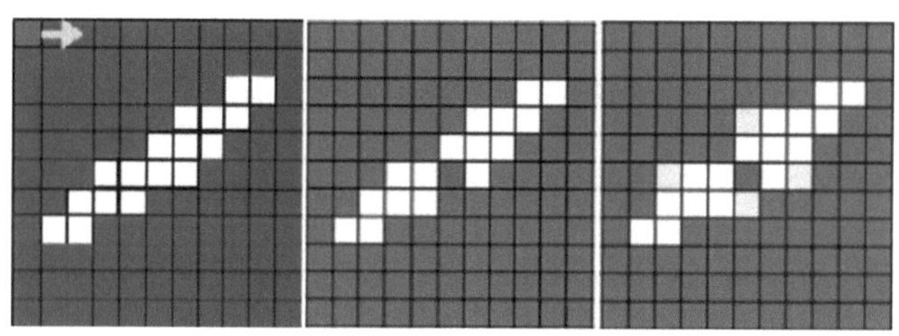

Fig. 6. Interpolation of pixel density for noise removal

3.3 Blotches and skin pores

The conversion of RGB images into L*a*b* space, noise removal, construction of a frequency image, measurement of the frequency distribution are the preprocessing steps needed for blotch and skin pore evaluation. In the following, the construction of the frequency image and measurement of the frequency distribution are detailed.

(1) Frequency image

A frequency image is an image in which the frequency of occurrence of the pixel value in the original image is shown, which is the color frequency information. Although a histogram is often used to represent frequency information in an image, the histogram has the disadvantage that the position information is lost in the process of computation. The frequency image has both the position and frequency information. The original and frequency images are shown in Fig. 7.

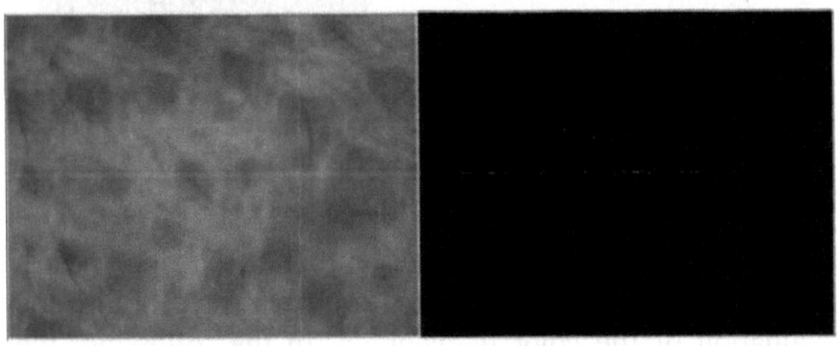

(a) original image (b) frequency image

Fig. 7. Original image in RGB color space and the resulting frequency image

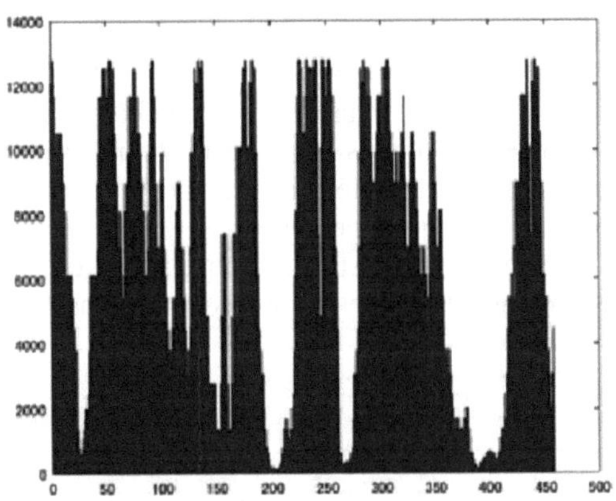

Fig. 8. Profile of the frequency image

(2) Frequency distribution measurement

In order to estimate the frequency image, we make a profile of the frequency image in the middle row of the image, as shown in Fig. 8.

The peaks and valleys of the graph correspond to the parts in which the color unevenness is high. When the valleys are deep and broad, the color unevenness is large and dark.

We use four thresholds at the frequency values of 5,000, 10, 000, 15, 000, and 20, 000 in Fig. 8. The number of peaks and valleys, width of the valleys, and the standard deviation of the valley width are computed at each threshold. When the frequency value is small and the above feature values are distinct, the color unevenness of an aggregated uneven region appears dominantly. The threshold value is empirically decided from the shape of profile image and the condition of the blotches and skin pores. In the subsequent analysis, we define the threshold value of 5,000 to be Level 1, 10,000 to be Level 2, 15,000 to be Level 3, and 20,000 to be Level 4. Although only one profile image in the middle of the image is demonstrated above, the same processing is performed over the entire region and the average values are computed for each index. The averaged indexes correspond to the number, area, and standard deviation of the area.

3.4 Discriminant analysis

A t-test at a 5% significance level was performed for the features obtained by skin grit, color unevenness, and skin pore processes. We then determined the effective features that have a significant difference between the index values in the cheek image and the chin image.

4 Results

The results of the test for the skin grit features are shown in Table 1, those of the color unevenness features are shown in Table 2, and those of the blotch and skin pore features are shown in Table 3. The values in the tables show the p values of the tests. In Table 1, the total number of the skin grit components, the average area of the grit components, and the standard deviation of their area are effective features for the assessment of skin grit. Table 2 shows that the maximum frequency value and standard deviation in the a* component images are effective for analysis, and that the standard deviation and kurtosis of the b* component images are also effective. Furthermore, Table 3 shows that the standard deviation of Level 1 in the a* component image as well as the average value and standard deviation of Levels 1, 3, and 4 are effective features.

Table 1. Results of skin texture feature discriminant analysis

	Total number	Mean area	Area std.
p value	1.3e-07	2.5e-04	3.9e-10

Table 2. Results of color unevenness feature discriminant analysis

(a) L* component image

	Max	Std.	Kur.	Skw.
p value	0.80	0.90	0.68	0.60

(b) a* component image

	Max	Std.	Kur.	Skw.
p value	7.05e-04	1.43e-02	5.70e-01	7.05e-01

(c) b* component image

	Max	Std.	Kur.	Skw.
p value	1.94e-01	8.90e-03	8.18e-07	3.07e-01

Table 3. Results of blotch and skin pore feature discriminant analy

(a) L* component image

	Number	Area mean	Std. area
Level 1	0.78	0.63	0.64
Level 2	0.85	0.81	0.80
Level 3	0.68	0.54	0.57
Level 4	0.43	0.09	0.16

(b) a* component image

	Number	Area mean	Std. area
Level 1	0.32	0.06	0.02
Level 2	0.67	0.53	0.64
Level 3	0.16	0.12	0.14
Level 4	0.46	0.11	0.25

(c) b* component image

	Number	Area mean	Std. area
Level 1	0.04	0.03	0.05
Level 2	0.20	0.42	0.26
Level 3	0.09	0.03	0.02
Level 4	0.08	0.01	0.01

5 Conclusion

In this study, we proposed an analysis method to extract skin texture and skin color unevenness features from magnified images of skin. In addition, we obtained effective features for the evaluation of skin condition by conducting two-sample t-tests on features extracted from images of the cheek and underside of the chin. As a result, for skin texture, we found that the total number of cristae cutis as well as the mean and standard deviation of those areas are effective features. For skin color unevenness, the maximum value of the a* image as well as the standard deviation and kurtosis of the b* image are effective. In addition, we created frequency images of colors for L*, a*, and b* images and analyzed the distribution and depth of the color. In this way, we obtainrd features for color unevenness, especially changes in colors, which correspond to blotches and skin pores. The test results show that the standard deviation of the area with color changes whose color frequencies are under 20,000 in an a* image and the mean and standard deviation of the area with color changes whose color frequencies are under 20,000, 10,000, and 5,000 in a b* image are effective.

References

1. Y. Baba, T. Mashita, Y. Mukaigawa, and Y. Yagi. Statistical Analysis of Human Skin Based on Reflection and Scattering Characteristics. IPSJ SIG Technical Report. 2009, vol. CVIM-167, no. 22 (2009)
2. H. Kobayashi, T. Hashimoto, K. Yamazaki, Y. Hirai. Proposal of Quantitative Index of Skin Texture by the Image Processing and Its Practical Application, Transactions of Japan Society of Mechanical Engineers Series C, vol. 76, no. 764, pp. 138-145 (2010) (in Japanese)
3. M. Nishioka. Measurement of Skin Texture Using Genetic Picture Analysis. The Institute of Electronics, Information and Communication Engineers. 2010-6, vol. 104, no. 140, pp. 65-69 (2010)
4. A. Sparavigna, R. Marazzato, An image processing analysis of skin textures, Skin Research and Technology, vol. 16, no. 2, pp. 161-167 (2010)
5. N. Ojima, I. Fujiwara, Y. Inoue, N. Tsumura, T. Nakaguchi, K. Iwata. Analysis on Unevenness of Skin Color Using the Melanin and Hemoglobin Components Separated by Independent Component Analysis of Skin Color Image, Proceedings of the SPIE, vol. 7897, 10.1117/12.873494

6. R. Rox Andescon. Polarized Light Examination and Photography of the Skin. Arch. Dermatol. 1991, vol. 7, no. 127, pp. 1000-1005 (1991)
7. N. Arakawa, H. Ohnishi, Y. Masuda. Development of Quantitative Analysis for the Micro-Relief of the Skin Surface Using a Video Microscope and Its Application to Examination of Skin Surface Texture, J. Soc. Cosmet. Chem. Japan, vol. 41, no. 3, pp. 173-180 (2007) (in Japanese)
8. N. Ojima. Image Analysis of Skin Color Using Independent Component Analysis and Its Application to Melanin Pigmentation Analysis, J. Soc. Cosmet. Chem. Japan, Vol.41, No.3, pp. 59-166 (2007) (in Japanese)
9. S. OE, T. KASHIWAGI , Imaging of Frequency Information of Color Image and its Applications, Journal of Institute of Image Electronics Engineers, vol.36, no.4, pp.530-535 (2007) (in Japanese)
10. Scalar Corporation, USB Microscope M3, http://www.scalar.co.jp/products/m3. php?cat=education

Effects of Photofrin-mediated photodynamic treatment on sensitivity to cisplatin in HeLa cells and the resistant subline

Yuichi Miyamoto[1], Yuri Yamaguchi[1] and Hatsuki Tajima[2]

[1] Department of Biomedical Engineering, Faculty of Health and Medical Care, Saitama Medical University, Saitama, Japan
ymiyamo@saitama-med.ac.jp
[2] Department of Clinical Engineering, Kimitsu Chuo Hospital, Kisarazu, Chiba, Japan

Abstract. Cisplatin (CDDP) is a widely used platinum-based compound that is effective against a broad spectrum of solid neoplasms. Although CDDP often leads to an initial therapeutic success, associated with partial responses or disease stabilization, many patients are intrinsically resistant to cisplatin-based therapies. The essential cytotoxic factors of photodynamic treatment (PDT) are reactive oxygen species (ROS). The reaction of ROS with cytoplasmic peptides and proteins may reduce the degree of inactivation of CDDP. In this study, we applied low-level PDT, which demonstrated no remarkable cytotoxic effects on CDDP–resistant HeLa cells (HeLa/CDDP) and examined whether PDT can reduce the CDDP-resistance level. There were no remarkable cytotoxic effects without CDDP in the PDT. The viability of the HeLa/CDDP-PDT group at a CDDP concentration of 10 μM was equal to that of normal HeLa cells. These findings suggest that PDT, which does not have sufficient cytotoxic effects, reduces the resistance level of HeLa/CDDP.

Keywords: Cisplatin, Photodynamic therapy, HeLa cells.

1 Introduction

Chemotherapy is one of the three most common types of cancer treatment alongside surgery and radiotherapy. The problem of acquired resistance to anticancer agents has been always a serious concern in addition to side effects. The mechanisms of resistance to anticancer agents in cells have been elucidated using mainly cancer cell lines. The presence of various mechanisms including reduction in drug uptake, enhancement in drug excretion, reduction in drug activation, drug inactivation, increase in target protein, decrease in the affinity to target protein, and enhancement in DNA repair has been shown in these cell lines [1]. Cisplatin (CDDP) is a platinum-based agent, whose efficacy has been demonstrated in a range of cancers, and it plays a central role in the current treatment with anticancer agents. Similar to other anticancer agents, continuous use of CDDP is known to lead to acquired resistance. In acquired CDDP-resistant cells, CDDP is inactivated through binding to metallothionein (a metal-binding protein) and glutathione (a peptide), within the cells [2].

Photodynamic therapy (PDT), a laser cancer treatment, has been established as a minimally invasive curative treatment for early superficial cancers [3]. PDT utilizes photochemical reactions leading to necrotic cancer cell death by oxidation, in which cancer patients are administered an oncotropic photopigment (photosensitizing agent) in advance. When the pigment is accumulated in a lesion, the pigment is excited, commonly by laser irradiation, to induce reactive oxygen species (ROS). We have sought to explore potential applications of PDT and previously reported the temporary activation of cell growth by PDT with low-level exciting red laser light, in addition to the enhancement in cytotoxicity of bleomycin combined with PDT [4, 5]. In this study, we built on our previous work in which we found that low-level PDT did not induce significant cell death and it regulates cellular function. We examined the effects of PDT on the reduction in resistance in HeLa cells with acquired CDDP-resistance (HeLa/CDDP cells), where a protein associated with the inactivation of CDDP in particular is present in excess levels.

2 Materials and Methods

We used HeLa cells derived from human cervical cancer (RIKEN CELL BANK). The culture medium was Ham's

J. Goh et al. (eds.), *The 16th International Conference on Biomedical Engineering*,
IFMBE Proceedings 61,
DOI: 10.1007/978-981-10-4220-1_3

F-10 medium (Invitrogen Co.) supplemented with 10% fetal bovine serum and 1% antibiotics (10,000 Units/ml Penicillin, 10,000 μg/l Streptomycin: GIBCO). For culturing HeLa/CDDP cells, CDDP (Wako Pure Chemical Industries, Ltd.) was added to Ham's F-10 medium. The cells were cultured at 37°C with 5% CO2−95% air. Photofrin (Pfizer Inc.) was used as a photosensitizer in PDT and cells were exposed to and rinsed in Ham's F-10 medium. The Cell Counting Kit-8 (Dojindo Molecular Technologies, Inc.) was used for the XTT viability assay. We used semiconductor laser light (Thorlabs, Inc. LP637-SF70, wavelength: 637 nm) and a laser driver (ASAHI DATA SYSTEMS, LD-ALP-7033CC) for laser excitation in PDT.

2.1 Preparation of HeLa/CDDP cells and evaluation of the degree of resistance

The growth inhibition curves of HeLa and HeLa/CDDP cells are shown in Figure 2. For both cells, cell viability tended to decrease as the CDDP concentration increased. The IC_{50} value, the concentration of CDDP at which growth inhibition is 50%, was calculated as 12.5 μM in HeLa cells, and 25.3 μM in HeLa/CDDP cells, resulting in a relative resistance (R.R) value of 2.1. We compared this with the results of Takara and colleagues, and judged that HeLa/CDDP cells had sufficient resistance to CDDP [6].

2.1 *CDDP cytotoxicity on photodynamic treated HeLa and HeLa/CDDP cells*

HeLa and HeLa/CDDP cells were seeded at 5×10^4 cells/100 μL in each well of a 96-well plate and cultured for 24 h. After this, the culture medium of each well was replaced with the Photofrin-containing (10 μg/mL) culture medium to allow the cells to be exposed for 15 min. The medium was then removed, the cells were rinsed with PBS(-), and the culture medium was replaced with the normal or CDDP-containing medium. The cells were irradiated with a continuous wave laser beam from below the 96-well plate (Fig. 1). The light dose ranged from 0.5 to 3.0 J/cm^2. The culture medium of samples following PDT was replaced with the normal or CDDP-containing medium and the cells were cultured for 48 h. After 10 μL of the XTT reagent was added into each well and the cells were incubated for 2 h, the absorbance was measured using a microplate reader to compare the cell viability between the groups.

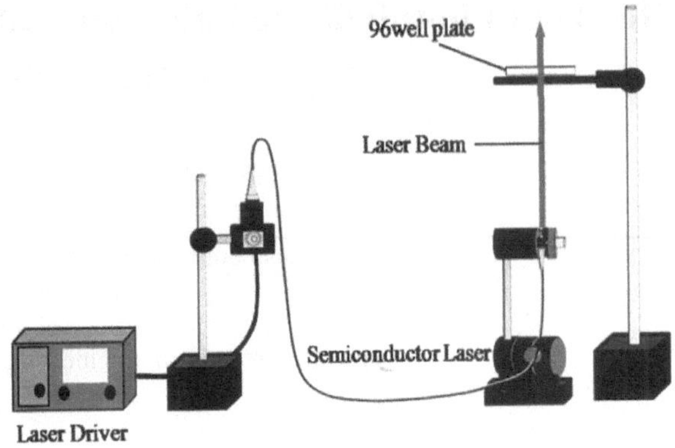

Fig. 1. Irradiation of HeLa and HeLa/CDDP cells with a laser beam. The test cells were cultured in a 96-well flat bottom plate. The cells were irradiated with a semiconductor laser beam (wavelength: 637 nm and power density: 5 mW/cm^2) from below the plate.

3 Results and Discussion

The growth inhibition curves of HeLa and HeLa/CDDP cells are shown in Figure 2. For both cells, cell viability tended to decrease as the CDDP concentration increased. The IC_{50} value, the concentration of CDDP at which growth inhibition is 50%, was calculated as 12.5 μM in HeLa cells, and 25.3 μM in HeLa/CDDP cells, resulting in a relative resistance (R.R) value of 2.1. We compared this with the results of Takara and colleagues, and judged that HeLa/CDDP cells had sufficient resistance to CDDP [6].

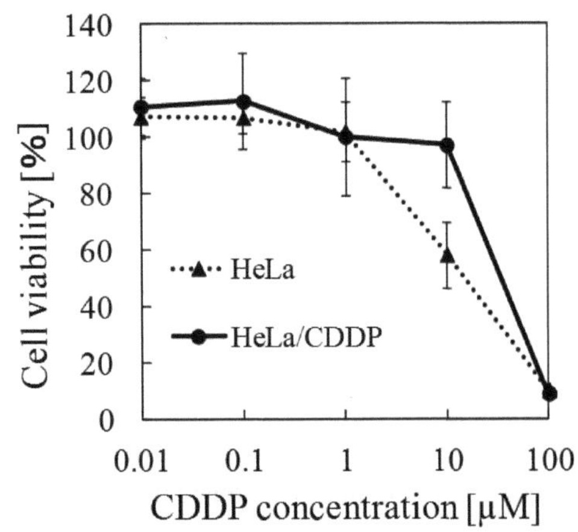

Fig. 2. Effect of CDDP concentration on HeLa and HeLa/CDDP cells. The growth activity at 48 h following the CDDP addition was measured. Data are shown as mean ± standard deviation, which was normalized by the viability of HeLa cells cultured in the normal medium (100%) (n = 4-6).

Table 1. Viability of HeLa and HeLa/CDDP cells in applied conditions. Data are shown as mean ± standard deviation, which were normalized by the viability of HeLa cells cultured in the normal medium (100%) (n = 4-6). The symbol (*) represents a significant difference in cell viability between the HeLa/CDDP group and HeLa/CDDP group treated with 10 μM CDDP at a light dose of 3.0 J/cm^2 (Student's t test : p <0.01).

PDT light dose [J/cm^2]	% Cell viability			
	HeLa	HeLa + CDDP (10 μM)	HeLa/CDDP	HeLa/CDDP + CDDP (10 μM)
0.0	100 ± 8.0	68.0 ± 2.0	100 ± 9.0	95.0 ± 9.0
0.5	83.0 ± 5.0	65.0 ± 3.0	92.0 ± 6.0	83.0 ± 11
1.0	74.0 ± 4.0	56.0 ± 5.0	94.0 ± 9.0	76.0 ± 7.0
3.0	44.0 ± 12	27.0 ± 5.0	92.0 ± 10	28.0 ± 8.0*

The cell viability values in the test groups are shown in Table 1. In Hela cells, the cell viability was kept ≥70% at a light dose of 0.5–1.0 J/cm^2, while it decreased to ≤45% at 3.0 J/cm^2. In the group where 10 μM CDDP, which is 80% of the IC_{50} value, was added into HeLa cells, the cell viability was kept ≥50% at a light dose of up to 1.0 J/cm^2, while it was below 30% at 3.0 J/cm^2, which shows a remarkable superimposed effect of PDT. In contrast, in HeLa/CDDP cells, the cell viability was kept ≥80% at a light dose of 3.0 J/cm^2 as well. Bi et al. showed that the intracellular amounts of ROS measured in acquired anticancer-resistant HeLa cells were approximately 3-fold higher than normal HeLa cells and suggested that acquired drug-resistant cells may have high resistance to ROS [7]. Once cells acquire resistance to an anticancer agent, they may become resistant to not only the anticancer agent, but also PDT, in which ROS are the major cytotoxic factors. In the group where 10 μM CDDP was added into HeLa/CDDP cells, the cell viability was kept ≥70% at a light dose of up to 1.0 J/cm^2, while it was ≤30% at 3.0 J/cm^2, which was nearly the same value as that observed in HeLa cells without resistance. In addition, this value is significantly lower than that in HeLa/CDDP cells, which indicates that the cytotoxicity of CDDP was effective in acquired CDDP-resistant HeLa cells and the degree of resistance is considered to be reduced by PDT.

Our findings in the present study suggest that ROS, cytotoxic factors in PDT, had some effects on the binding activity of metallothionein and glutathione, both of which are largely associated with the inactivation of CDDP in acquired CDDP-resistant cells.

Acknowledgments

This study was supported by the JSPS KAKENHI Grant Number 15K01296. We thank Mr. Yuta Sekimoto for cooperation in conducting this study.

References

1. Naito, S., Yokomizo, A., Koga, H.: Mechanisms of drug resistance in chemotherapy for urogenital carcinoma. Int J. Urol. 6, 427–439 (1999)
2. Galluzzi, L., Senovilla, L., Vitale, I., Michels, I., Kepp, O., Castedo, M., Kroemer, G.: Molecular mechanisms of cisplatin resistance. Oncogene 31(15), 1869-1883 (2012)
3. Dougherty, T. J.: Photodynamic therapy. Photochem. Photobiol. 58, 895-900 (1993)
4. Miyamoto, Y., Nishikiori, D., Hagino, F., Wakita, M., Tanabe, I., Toida, M.: Effect of 630-nm pulsed laser irradiation on the proliferation of HeLa cells in Photofrin-mediated photodynamic therapy. Laser Therapy 20(2), 135-138 (2011)
5. Nishikiori, D., Miyamoto, Y.: Enhancement of the cytotoxic effects of bleomycin with permeabilization of the plasma membrane by Photofrin-mediated photodynamic therapy in vitro. IFMBE Proceedings 43, 711-713 (2013)
6. Takara, K., Obata, Y., Yoshikawa, E., Kitada, N., Sakaeda, T., Ohnishi, N., Yokoyama, T.: Molecular changes to HeLa cells on continuous exposure to cisplatin or paclitaxel. Cancer Chemother Pharmacol. 58, 785-793 (2006)
7. Bi, W., Wang, Y., Sun, G., Zhang, X., Wei, Y., Li, L., Wang, X.: Paclitaxel-resistant HeLa cells have upregulated levels of reactive oxygen species and increased expression of taxol resistance gene 1. Pak J Pharm Sci. 27(4), 871-878 (2014)

Novel spectral approach for pulse waves contour analysis

A.A. Fedotov[1], A.S. Akulova[2], S.A. Akulov[1], M.A. Reshetnikova[3]

[1] Samara State Aerospace University, Department of Laser Systems and Biomedical Engineering, Associate Professor, Samara, Russia
[2] Samara State Aerospace University, Department of Laser Systems and Biomedical Engineering, Ph.D. student, Samara, Russia
[3] Samara State Aerospace University, Department of Laser Systems and Biomedical Engineering, graduate student, Samara, Russia

Abstract— **This paper considers novel techniques for pulse wave contour analysis. Classical morphological pulse wave analysis by using stiffness index and reflection index does not provide the required accuracy of diagnosis for patients with severe arterial stiffness due to indistinguishable dicrotic peak of pulse wave. We proposed a novel approach for pulse wave contour analysis based on spectral analysis of replicated single pulse waveform. The newly developed spectral harmonic index was defined as the amplitude ratio of the first two harmonics of the amplitude spectrum. Comparative analysis of sensitivity and specificity for various indices for a set of 45 volunteers of different age and arterial stiffness was carried out. It was found that the proposed spectral harmonic index has the highest values of sensitivity (89%) and specificity (86%) compared to the existent indices defined in the time domain.**

Keywords— **digital pulse wave, contour analysis, spectral transform, arterial stiffness.**

I. INTRODUCTION

Early diagnosis of human cardiovascular pathologies may use methods based on estimation of arterial stiffness for the purpose of prognostic evaluation of atherosclerosis, coronary heart disease and other serious cardiac diseases [1]. Modern clinical approaches and instruments for noninvasive diagnosis of blood vessels includes noninvasive pulse wave recording (most common by using photoplethysmography technique) with further application of advanced signal processing methods [2]. Estimation of arterial stiffness could be realized by contour analysis of the pulse waveform [3].

The shape of pulse waves, propagating along blood vessels, is mainly formed by the interaction between the left ventricle of the heart and large vessels of the systemic circulation, and reflects the composition of forward and reflected pulse waves [4]. The aim of this work is to develop new approach as well as analysis of existing methods for pulse wave contour analysis in order to determine the most effective noninvasive tool.

II. THEORY

The composition of the digital pulse waveform is given schematically in Fig. 1; where ➤ is the forward pulse wave; ⊳ is the reflected pulse wave; and • is the femoral bifurcation of the aorta; A_1 – amplitude of the forward wave, A_2 – amplitude of the reflected wave (amplitude of dicrotic peak); Δ – the time of the pulse wave reflection.

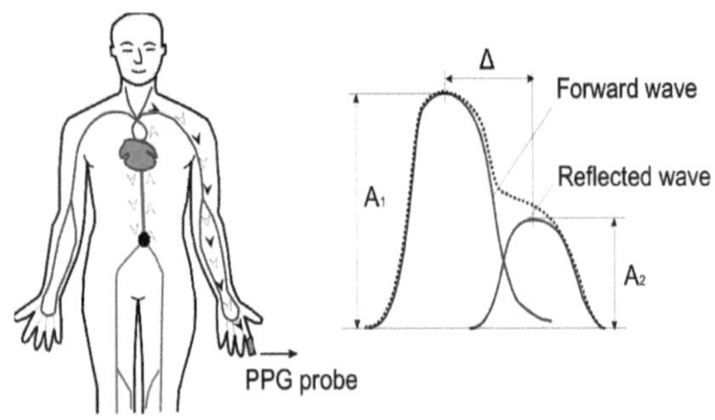

Fig. 1. The composition of digital pulse waveform

The forward pulse wave is formed by the systolic blood volume, which transferred directly from the left ventricle to the arteries of the upper extremities during the systole. The reflected pulse wave arises mainly from pressure waves transmitted along the aorta to small arteries in the lower extremities, where they are then reflected back along the ascending aorta and then travels to the finger arterioles [3, 4].

The intensity of the reflection (amplitude of the reflected wave A_2) is determined by the tone of the small muscle arteries. The time of the pulse wave reflection (Δ) depends primarily on the arterial stiffness and pulse wave velocity in the following segment of the arterial bed: *aorta – large arteries – femoral bifurcation of the aorta.*

The pulse wave velocity depends on the elasticity of blood vessels, their geometry and blood density and is determined by the Moens-Korteweg equation [4]:

$$v = \sqrt{\frac{Eh}{2\rho r}}$$

J. Goh et al. (eds.), *The 16th International Conference on Biomedical Engineering*,
IFMBE Proceedings 61,
DOI: 10.1007/978-981-10-4220-1_4

where: v – pulse wave velocity; E – the Young's modulus of the arterial wall; h – the arterial wall thickness; ρ – blood density; r – the arterial radius.

The basic hemodynamic parameters that affect the pulse wave velocity, and hence, the pulse waveform is arterial stiffness, which in turn, is an effective and reliable prognostic marker of atherosclerosis and endothelial dysfunction [5]. Thus it is undeniably very important to create novel approach for pulse wave contour analysis in order to estimate the arterial stiffness by using simple, reliable, noninvasive and inexpensive diagnostic methods based on recording and digital processing of arterial pulse waves.

III. Materials and Methods

Currently in clinical practice for the contour analysis of digital pulse waves the reflection index (RI) and the stiffness index (SI) are widely used. These indices are defined as follows:

$$RI = \frac{A_2}{A_1} \cdot 100\%, \quad SI = \frac{L}{\Delta}$$

where: L is the subject's height expressed in meters.

The time of the pulse wave reflection Δ estimates arterial stiffness and also depends on the length of corresponding arterial bed. The variation of arterial path length with a patient's height could be addressed by means of a stiffness index defined as height divided by the time of the pulse wave reflection. The reflection index RI is determined by the intensity of pulse wave reflections and depends on the tone of small muscular arteries and impedance mismatches in arterial system [1, 3].

However using traditional indices for contour analysis of the pulse wave for patients with severe arterial stiffness is associated with a number of obstacles and difficulties, the main of which is the complication of detecting the reflected wave maximum (so called dicrotic peak) and the further definition of its amplitude and temporal position [6]. Below you may see the visual confirmation of this well-known phenomenon.

As an alternative index for the pulse wave contour analysis we could propose to use the form factor of pulse wave defined as follows [8]:

$$FF = \frac{\sigma_{x''} / \sigma_{x'}}{\sigma_{x'} / \sigma_x}$$

where: FF is the form factor; $\sigma_{x''}$ is the standard deviation of the second derivative of the considered pulse waveform; $\sigma_{x'}$ is the standard deviation of the first derivative of the considered pulse waveform; σ_x is a standard deviation of the considered pulse waveform.

The transition from the time domain to the frequency domain by using the spectral Fourier transform allows to obtain a clearer assessment of the pulse wave morphological characteristics. Given the fact that in practice the pulse wave signal processing is realized by software, the transition to the frequency domain is implemented by means of discrete Fourier transforms. In this case the expression for the amplitude spectrum of the pulse wave could be determined as follows:

$$X(k) = \sum_{n=0}^{N-1} x(n) \cdot e^{-j\frac{2\pi}{N}nk}$$

where: $x(n)$ is a sample of the pulse wave in time domain; n is sample's number in the time domain; k is sample's number in the frequency domain; N is a total number of samples in the considered sequence of pulse waves; $X(k)$ is a sample of the pulse wave in the frequency domain.

To obtain novel index for pulse wave morphology assessment we perform Fast Fourier Transform for the sequence formed by consecutive replication of a single pulse waveform with removed dc component. For accurate extraction of single pulse waveform from raw biosignal sequence the adaptive detector of pule wave beats was used [7]. This detector is characterized by small errors in detecting maximums and onsets of pulse waves, contaminated by various types of artifacts and noise of different origin and intensity.

Analysis of the obtained spectral characteristics for different types of pulse waveforms has shown significant differences in the structure of the amplitude spectrum. For numerical evaluation of these spectral differences we suggest the use of spectral harmonic index (SHI) defined as the amplitude ratio of the first harmonic (As_1) to the second harmonic (As_2):

$$SHI = \frac{As_1}{As_2}$$

For practical studies of different approaches to the pulse wave contour analysis it is necessary to collect clinical pulse wave recordings from people with various conditions of arterial vessels. The pulse wave signals were recorded by using certified photoplethysmography device ELDAR ("New Devices Ltd", Samara, Russian Federation) with the probe located at the cuticle of forefinger of the right hand, sampling rate of 100 Hz, ADC resolution of 10 bits, bandwidth of 0,05 – 15 Hz.

Three groups of volunteers, consisting of 75 people aged from 20 to 70 years were examined. The 1st group (Group A) consisted of 30 young healthy people aged 20 to 35 years; the 2nd group (Group B) consisted of 25 people at ages from 40 to 55 years with typical age-related changes in arterial stiffness, but without significant cardiac diseases; and the 3rd group (Group C) consisted of 20 elderly people with different cardiovascular diseases aged 60 to 75 years. Informed consent was obtained from all participants. The Ethical Committee of the Regional Hospital at Samara has approved the clinical study.

The study was conducted in a regional clinical hospital in a quiet room at 22^0 C. Pulse wave recording for all subjects was performed in the period from 8 a.m. to 10 a.m. on an empty stomach after a 10 minute rest in the supine position. All drugs were canceled 12 hours before the study. To assess differences between mean values in compared groups we used the non-parametric Mann-Whitney U test. All experimental data are presented as M ± SD (M is the mean value of the arterial index and SD is the standard deviation), the differences were considered significant at $p < 0{,}02$.

IV. Results

Fig. 2 shows typical pulse waveforms in the time domain for each of the three groups, the amplitude of the pulse wave is given in arbitrary units. Typical waveforms from each group demonstrate obvious morphological features of the pulse wave signal and emphasize distinctions among the considered groups. Figure 3 illustrates the amplitude spectrum of the corresponding replicated pulse waveforms from Figure 2.

With developing severe functional changes in the arterial system (see Fig. 3), the bimodal shape of pulse wave disappears, and the dicrotic peak becomes indistinguishable, which leads to significant difficulties in using classic indices for pulse wave contour analysis.

Table 1. Arterial indices for three observed groups

Group	Arterial indices M ± SD			
	SHI, units	*FF, units*	*SI, m/s*	*RI, %*
Group A	1.05±0,16	1.91±0,22	7.5±0,65	54±11
Group B	1.48±0,36	1.51±0,31	9.8±1,15	61±15
Group C	2.25±0,34	1.25±0,19	n/a	n/a

The values of different arterial indices for three observed groups are listed in Table 1 (n/a means non applicable).

Table 2 contains the results of quantifying the statistical reliability of different indices for arterial stiffness assessment; we have calculated Specificity (S^+) for two groups of healthy people (*Groups A* and *B*) and Sensitivity (S^-) for the group of people with cardiovascular diseases (*Group C*) [9].

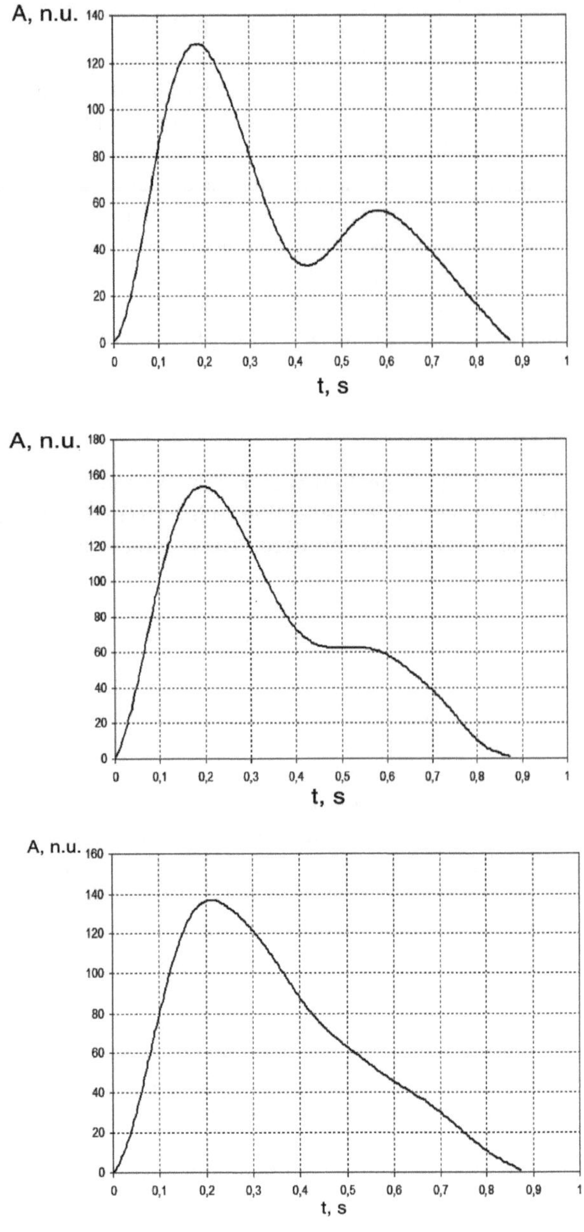

Fig. 2. Typical pulse waveforms: top trace is from group A; middle trace is from group B; bottom trace is from group C

We were unable to obtain any reliable results of calculating *SI* and *RI* for the participants from *group C* due to inevitable issue with dicrotic peak detection, at the same time there were no problems with defining arterial indices for that group by using proposed spectral approach or calculating form factor of single pulse waveform. Obtained results also suggest that there are substantial and significant differences ($p < 0{,}02$) in the values of form factor *FF* and the *SHI* defined for people with different functional state of the arterial system.

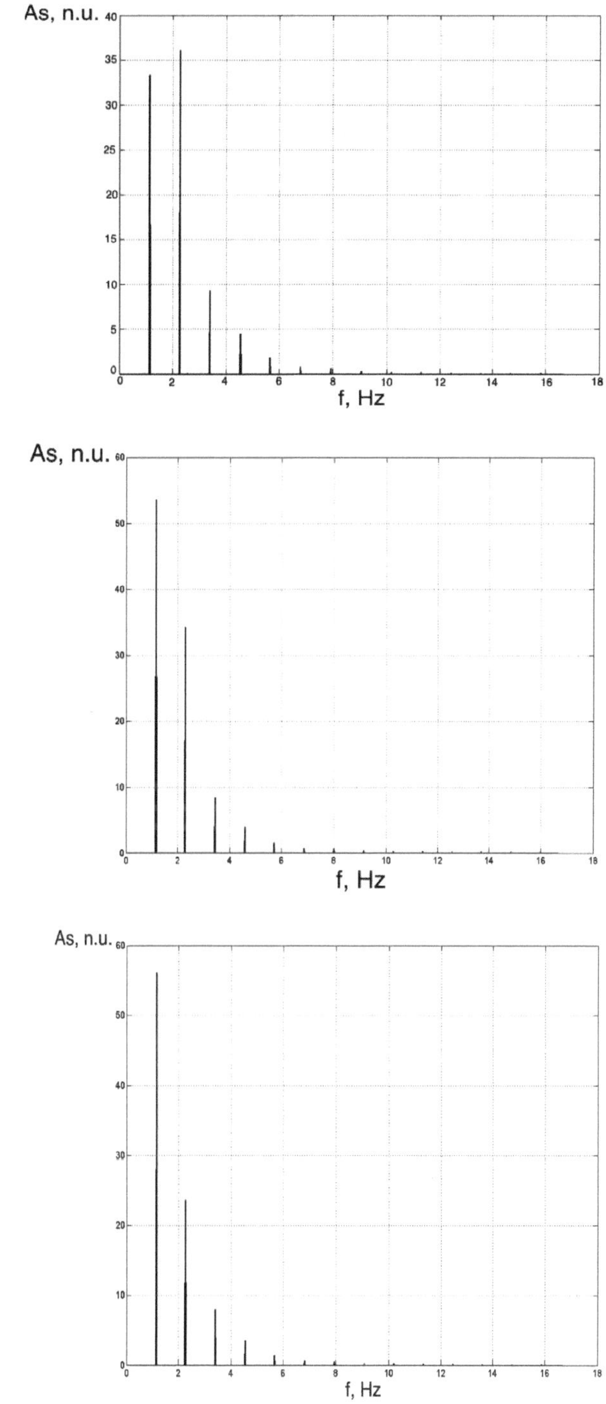

Fig. 3. The amplitude spectrum for replicated pulse waveform from Fig. 2; top trace is from group A; middle trace is from group B; bottom trace is from group C

Table 2. The sensitivity and specificity for different indices of pulse wave contour analysis

Group	Sensitivity S^+, %				Specificity S^-, %			
	SHI	*FF*	*SI*	*RI*	*SHI*	*FF*	*SI*	*RI*
Group A	–	–	–	–	**86**	82	78	72
Group B	–	–	–	–	**85**	81	78	74
Group C	**88**	80	n/a	n/a	–	–	–	–

V. Conclusion

The results of our study indicate that using spectral analysis for sequence of replicated pulse waveform is a powerful method for noninvasive arterial vessels diagnostics, which characterized by higher values of sensitivity and specificity compared to classic contour analysis implemented in the time domain. Spectral analysis of replicated pulse waveform could be an effective basis for creating novel diagnostic algorithms for assessment of arterial stiffness based on the pulse wave contour analysis.

The clear advantage of using the proposed spectral approach in comparison with classic time domain methods is that there is no need for accurate detection of the pulse wave's dicrotic peak that is crucial in case of assessment arterial stiffness for elderly people with severe cardiac diseases and almost indistinguishable dicrotic peak. The disadvantage of the suggested technique may include the need for correct extraction of the single pulse wave, which in practice may be difficult due to present baseline wander and movement artifacts, heavily distorting the shape of the pulse waveform.

Conflict of Interest

The authors declare that they have no conflict of interest.

References

1. Expert Consensus Document on arterial stiffness: methodological issues and clinical applications. (2006). European Heart Journal 27 (21): 2588–2605
2. Allen J. (2007). Photoplethysmography and its application in clinical physiological measurement. Physiological Measurement 28: 1-39
3. Millasseau S.C. et al. (2006). Contour analysis of the photoplethysmographic pulse measured at the finger. Hypertension 8: 1449–1456
4. Weber T. (2004). Arterial Stiffness, Wave Reflections, and the Risk of Coronary Artery Disease. Circulation 109: 184–189

5. Millasseau S.C. et al. (2002). Determination of age-related increases in large artery stiffness by digital pulse contour analysis. Clinical Science 103: 371–377
6. Fedotov A.A. (2015) A mathematical model of hemodynamic processes for distal pulse-wave formation. Biophysics 60 (2): 274-277
7. Fedotov A.A. (2013) Amplitude–Time Method for Detecting Characteristic Pulse Wave Points," Biomedical Engineering 46 (6): 241-245
8. Rangayyan R.M. (2002) Biomedical Signal Analysis: A Case-Study Approach. IEEE Press and Wiley, New York
9. Altman D.G., Bland J.M. (1994) Diagnostic tests 1: sensitivity and specificity. British Medical Journal 308: 1552

Corresponding author:

Author: Fedotov Aleksandr Aleksandrovich
Institute: Samara State Aerospace University
Street: Kirova avenue, 303, #11
City: Samara
Country: Russian Federation
Email: fedoaleks@yandex.ru

A Comprehensive Analysis on Breast Cancer Classification with Radial Basis Function and Gaussian Mixture Model

Harikumar Rajaguru, Sunil Kumar Prabhakar

Department of ECE, Bannari Amman Institute of Technology, India

Abstract— Cancer is a type of deadly disease where a particular group of cells display growth which becomes uncontrollable after a certain period of time. Breast Cancer is a type of cancer which affects the inner lining of the milk ducts. The symptoms of breast cancer include the shape alteration of the breast size, nipple discharge, swelling of the lymph node and pain in the nipple. There are several types of breast cancer like lobular carcinoma, ductal carcinoma, invasive lobular carcinoma, inflammatory breast carcinoma etc. The risk factor of breast cancer includes factors like sex, hormonal fluctuations, alcohol intake, environmental and genetic factors, other abnormalities in the human body along with a high fat diet. In this work, a simple, cost effective and non-invasive strategy to detect the breast cancer at an early stage is proposed with the help of techniques such as Gaussian Mixture Model (GMM) and Radial Basis Function (RBF). As cancer staging is divided into clinical and pathological stage, the TNM (Tumour Node Metasis) prognostic tools are identified and the TNM variables such as tumour size, history of breast feeding, menstrual cycle, hereditary, food habits, etc. are used as input variables for both the types of classifications. The data collection was obtained from the cancer centre of Kuppuswamy Naidu Memorial Hospital, Coimbatore, India. The Performance Metrics taken here are Specificity, Sensitivity, Accuracy, Perfect Classification, Missed Classification, False Alarm and Performance Index. Results show an average accuracy of 89.60% is obtained with GMM classifier and an average accuracy of 92.75% is obtained with RBF classifier.

Keywords— Breast Cancer, RBF, GMM, TNM, Accuracy

1 Introduction

One of the biggest threats to human life is cancer and today more women in the world are affected by the Breast Cancer [1]. It has become the need of the hour and a big challenge to fight against cancers from a medical and scientific point of view. Cancer in the early stage is somewhat easy to treat whereas cancer is the most advanced stages are absolutely difficult to treat. The staging of any form of cancer is divided into two stages, namely, clinical and pathological stage [2]. Cancer is a group of disease which involves the cell growth in an abnormal manner with the potential to severely spread to various other organs of the body. All the tumours need not be cancerous and benign tumours cannot spread fastly to other sections of the body. The most common symptoms include abnormal bleeding, a lump, prolonged cough, change in bowel movements, and an unexplained weight loss. In the TNM stage (Tumour, Node, and Metasis) stage, prognostic tool are traced and new innovative techniques for development of prognostic tools have been developed [3]. With the help of pre-operative tests which include mammography, core needle biopsy, fine- needle aspiration cytology, accompanied by various other types of physical examinations, then the diagnosis of the palpable breast lesions can be accurately done. For screening purposes, mammography is quite often used hence it gives the maximum possibilities for a physician to trace the exact location of micro calcifications and other possible indicators in the tissue of breast.

The patient is sent for further consultation to the pathologist if a suspicious region is found. Due to its low cost, easy nature and fast approach, fine needle aspiration technique is used as it allows the pathologist to examine the breast tissue more closely. Several automated works are proposed in the literature as follows. The machine learning techniques to diagnose breast cancer from image-processed nuclear features of fine needle aspirates was done by William et al [4]. Lukasz et al showed the classification of breast cancer malignancy using cytological images of fine needle aspiration biopsies [5]. Nithya and Santhi did a comparitive study on feature extraction method for breast cancer classification [6]. The feature selection for breast cancer malignancy classification problem was reported by Filipczuk et al [7]. The system for remote cytological diagnosis and prognosis of breast cancer was explained by William N.Street [8]. Breast mass classification based on cytological patterns using RBFNN and SVM was reported by Subashini et al [9]. The support vector machines combined with feature selection for breast cancer diagnosis was done by Akay [10]. Ensemble strategies for a medical diagnostic decision support system for a breast cancer diagnosis application was done by West et al [11]. The multisurface method of pattern separation for medical diagnosis applied to

J. Goh et al. (eds.), *The 16th International Conference on Biomedical Engineering*,
IFMBE Proceedings 61,
DOI: 10.1007/978-981-10-4220-1_5

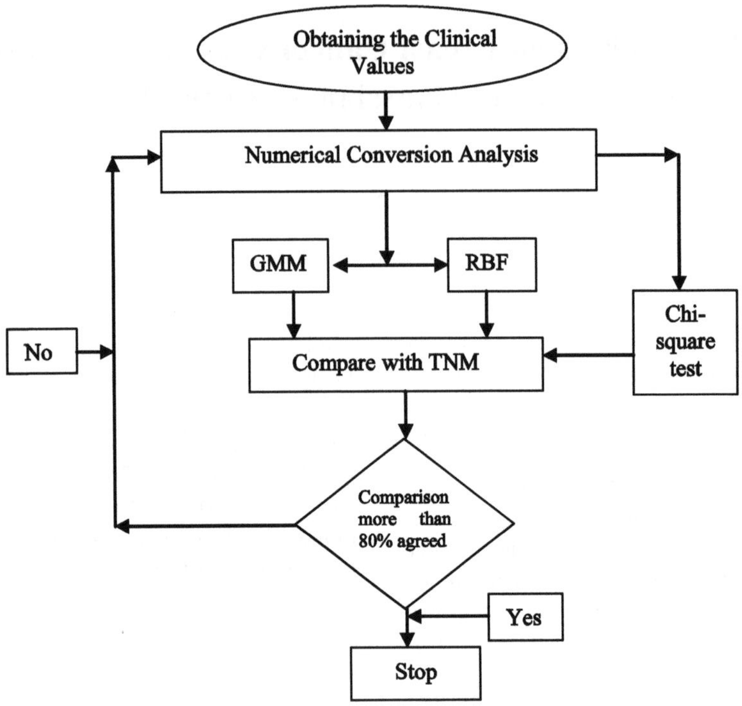

Figure 1 Block Diagram of the Work

breast cytology is done by Wolberg and Mangasarian [12]. The breast cancer prediction using the isotonic separation technique is given by Ryu et.al [13]. A new hybrid method based on fuzzy- artificial immune system and K-nn algorithm for breast cancer diagnosis is given by Sahan et al [14]. Ubeyli implemented an automated diagnostic system for breast cancer detection [15]. In this paper, the classification accuracy of the TNM staging process is compared with the Chi-square test, RBF and GMM. In this study, the total numbers of patients with breast cancer studied are 82. For all the types of classifications, the input variables are nothing but the TNM variables (such as distance Metasis, tumour size, number of positive and negative nodes, history of breast feeding, hereditary, menstrual cycle etc...) It is observed that the Chi-Square classification are pretty close to the investigations done in a clinical manner. Artificial Neural Network (ANN) such as RBF is more accurate and GMM is also one of the best measure to classify the breast cancer. To increase the classification accuracy,new prognostic factors can be added to the ANN. The block diagram of the work is shown in Figure 1. The paper structure is as follows. In section 2, the materials and methods are discussed followed by the application of GMM, RBF and Chi-Square Test for the correct classification in section 3. Section 4 deals with the results and discussion followed by the conclusion in Section 5.

2 Materials and Methods

The figure 1 clearly depicts the entire process. Initially the meta data , i.e the clinical values are converted into a numerical form of values. The obtained numerical values has to undergo, Chi-square test. The values obtained through the Chi-square test is then compared with the various TNM stages of classifications. The breast cancer stages are classified with the help of RBF and GMM. Totally 82 patients with breast cancer are taken for the study. The data was obtained from the Oncology Department from Kuppuswamy Naidu Memorial Hospital, Coimbatore. This study analysis was dependent on the consolidated analysis obtained in the referral letters, hospital charts, operative reports, radiological studies, radiation therapy and pathological reports. The cancer patient were clearly sorted and analyzed based on the parameters like general habits, food habits, marital status, family history cancer, age of menarche, menstrual cycle, menstrual duration, number of abortions, menopause, family planning, Breast feeding factors, overall health and hygiene, IDL-DL scope, T-stage, Breast size,

Breast nipple position, Breast nipple level, lump location, lump size, lump T-stage, CVS, CNS, RS, and Abdomen readings. Based on the standard 1992 classification of the UICC (International Union Against Cancer), the clinical staging of tumours are done. The stage distribution according to the UICC classification is as follows. In stage I, 20 patients are present, in stage II 28 patients are present, in stage III 11 patients are present and in stage IV 23 patients are utilized as controlled patients for the validation and classification purposes. Chi-Square test is used to verify the usage of classification of Breast Cancer Classification.

2.1 Chi-Square Test:

The first step in Chi-square test is that the null hypothesis should be formulated. As the clustering process is highly random in nature and inconsistent, the significant level α is selected. The value of χ^2 is computed as

$$\chi^2 = \frac{\Sigma(f_0 - f_e)^2}{f_e}$$

where the observed frequency is denoted as f_0 and the expected frequency is denoted as f_e.

The distribution quantity χ^2 is carefully examined. The null hypothesis is rejected as $\chi^2 > \chi^2\ \alpha_0;\ df$ where $\chi^2\ \alpha_0;\ df$ gives the significance level α_0 and df denotes the degree of freedom. The degree of freedom represents the total number of independent frequencies. The values of Chi square were done from the T tests and F tests respectively for the 26 variables on the 82 cases and the results are tabulated in Table 1.

Table 1 Performance of Chi Square Test for the Breast Cancer Patients

	Performance of Chi-Square Test		
	Minimum Value	Maximum Value	Average Value
T1	3.84	19.52	11.7
T2	21.2	32.96	26.9
T3	30.72	40.85	35.5
T4	41.92	90.08	66

Based on the T test and F test, the values of Chi-square are tabulated for 26 variables belonging to the 82 patients. The T1 stage showed a particular result of 3.84-19.52 and it shows the initial prognosis stage of Tumour 1. The T4 stage showed a particular result of 41.92-90.08, which is a very dangerous state and invasive stage of breast cancer. The classification accuracy of Chi-square test in terms of its efficiency is compared to the clinical data. The observations show that the values obtained through Chi-square test is close to the values obtained clinically and so this technique can be used to detect the classification of cancer very earlier. With the advent of various artificial neural networks tools and bioinformatics tools, further decision making of classification of breast cancer can be done more accurately.

3 Classifiers Used Here for Breast Cancer Classification

The two important post classifiers used here are Radial Basis Function and Gaussian Mixture Model

3.1 Radial Basis Function:

One of the non-linear and statistical tools for modeling the data is neural network and it is utilized for modeling very complex relationships in between the inputs and outputs. To seek the different patterns in a dataset, RBF is used mostly [16]. This type of neural networks comes under the category of feed forward neural network, comprising of three layers such as input layers, hidden layer and output layer. Different tasks are performed by each layer. Determination of the outputs of the input layer in RBF networks is important and it is done by calculation of the respective distances between the network inputs and the centre of the hidden layer. The second layer is called the hidden layer which is linear in nature and the outputs of this hidden layer are weighted forms of the outputs of the input layer. In hidden layer, each neuron has a vector parameter representation called centre. The expression is as follows

$$\hat{z}_j = \sum_{p=1}^{P} m_{ps}\phi\left(\left\|f - d_p\right\|\right) + \beta_s$$

Euclidean distance is the norm assumed here and Gaussian function is the radial basis function assumed here and is defined as

$$\varphi(r) = \exp\left(-\alpha_p \cdot \left\|f - d_p\right\|^2\right)$$

where the total number of neurons in the hidden layer is represented by P and the total number of neurons in the output layer is represented as S . The weight factor between the pth neuron and the sth output is given by the term m_{ps}. The radial basis function is represented as φ. The respective spread parameter of the pth neuron is α_p. The input data vector is denoted as f, the centre vector of the pth neuron is represented as d_p. The bias value of the output s^{th} neuron is represented as β_s. The network output of the sth neuron is represented as $\hat{z}_S$.

In RBF networks, L dimensional inputs (f_1,...., f_q) is present in the input layer, which later broadcasts the inputs to the hidden layer. There are totally P neurons in the hidden layer and the Euclidean distances between the inputs and centres are calculated in this layer. The function which is nothing but an activation function is present in the hidden layer. The usually chosen radial basis function is Gaussian function and to shape the curve (α_1,...., α_p) it has a spread parameter. The (m_{11},...., m_{ps}) outputs from the hidden layer are then transferred to the output layer. Here $P(p \in \{1,2,...P\})$ represents the total number of neurons in the hidden layer and $S(s \in \{1,2,....S\})$ represents the dimensions of the output. The linear combination of the outputs is in the hidden layer along with the bias parameters is calculated by output layer and at the end, the output of the RBF networks is obtained $(\hat{z}_1,....,\hat{z}_S)$.

To decide the total number of neurons in the hidden layers, the most important procedure in the design of RBF network. So to get the desired output of the RBF network m,α,d,β parameters should be properly adjusted. MSE is the standard error metrics reference to evaluate the performance of the network. Error expression of the RBF network is as follows

$$E^{SSE(m,\alpha,d,\beta)} = \sum_{s=1}^{S}(y_s - \hat{y}_s)^2$$

Here the derived output is denoted as y_s and $\hat{y}_S$ denotes the output of the RBF neural network. Training has to be carried out in the RBF network so that the error function is minimized.

KF algorithm is used here as a training algorithm. A finite and non linear dimensional system has a discrete time is represented as

$$\theta_{e+1} = \theta_e + w_e$$
$$z_e = h(\theta_e) + t_e$$

where θ_e denotes the state of the system at a time instant e, w_e represents the noise process, z_e represents the observation vector and t_e represents the observation noise and h represents the vector function of the state which is non linear in nature. As a least squares minimization problem, the Kalman Filtering algorithm is utilized to optimize both the centre vectors and weight factors. In RBF networks, the target output vector is denoted as

$$z = [z_1,...,z_S]^T$$
$$h(\theta_e) = [\hat{z}_1,.....,\hat{z}_S]^T$$

The vector θ is having all RBF parameters as

$$\theta = [w_{11},....,w_{1S},d_1,....,d_P)^T$$

3.2 Gaussian Mixture Model:

The univariate Gaussian distribution is represented mathematically as follows, $N(x|\mu,\sigma) = \frac{1}{\sqrt{2\pi\sigma^2}} e^{-\frac{(x-\mu)^2}{2\sigma^2}}$, where μ represents the mean and σ represents the variance. The multivariate Gaussian distribution is represented as follows

$$N(x|\mu,\Sigma) = \frac{1}{(2\pi|\Sigma|)^{1/2}} \exp\left\{-\frac{1}{2}(x-\mu)^T \Sigma^{-1}(x-\mu)\right\}$$

μ represents the mean and S represents the covariance.

The estimation of it can be done by Maximum Likelihood estimation or Expectation Maximization Algorithm. The log of Gaussian distribution is considered as

$$\ln p(x|\mu,\Sigma) = -\frac{1}{2}\ln(2\pi) - \frac{1}{2}\ln|\Sigma| - \frac{1}{2}(x-\mu)^T \Sigma^{-1}(x-\mu)$$

The derivative is obtained and equated to zero and is expressed as follows

$$\frac{\partial \ln p(x|\mu,\Sigma)}{\partial \mu} = 0 \text{ and } \frac{\partial \ln p(x|\mu,\Sigma)}{\partial \Sigma} = 0$$

The above equations imply the following

$$\mu_{ML} = \frac{1}{N}\sum_{n=1}^{N} x_n \text{ and } \Sigma_{ML} = \frac{1}{N}\sum_{n=1}^{N}(x_n - \mu_{ML})(x_n - \mu_{ML})^T$$

where N represents the number of data points or samples The linear position of Gaussians is represented as follows [17]

$$p(x) = \sum_{k=1}^{K} \pi_k N(x|\mu_k,\Sigma_k)$$

Where K represents the number of Gaussians and π_k represents the mixing coefficient which represent the weightage for each Gaussian district.

The normalization process should satisfy the following condition as

$$0 \le \pi_k \le 1, \sum_{k=1}^{K} \pi_k = 1$$

If the log likelihood function is considered as follows

$$\ln p(X|\mu,\Sigma,\pi) = \sum_{n=1}^{N} \ln p(x_n) = \sum_{n=1}^{N} \ln\left\{\sum_{k=1}^{K} \pi_k N(x_n|\mu_k,\Sigma_k)\right\}$$

As there is no closed form of solution here, ML does not work here and so the parameters are calculated only based on the EM algorithm.

The mixing coefficients are assumed as prior probabilities for the components. For a particular value ofx', the posterior probabilities also called as responsibilities are evaluated easily. From the Bayesian rule , the latent variable is given as

$$\gamma_k(x) = p(k|x) = \frac{p(k)p(x|k)}{p(x)} = \frac{\pi_k N(x|\mu_k,\Sigma_k)}{\sum_{j=1}^{K} \pi_j N(x|\mu_j,\Sigma_j)},$$

where $\pi_k = \frac{N_k}{N}$

One of the optimization technique operated iteratively and in a local manner is EM algorithm. It consists of two steps namely E-step and M-step.

E-step: The expected values of the latent variable for a set of parameter values can be computed easily in this stage

M-step: Using the latent variables which are calculated from ML method, the parameters of the model is updated.

3.2.1 EM Algorithm for GMM:

The main goal is to maximize the likelihood function based on the parameters which consists of mean and covariance only

Step :1 The mean μ_j, covariance Σ_j. and the mixing coefficients π_j. are initialized and then the initial value of the log likelihood is initialized.

Step :2 The responsibilities using the present parameter values are evaluated as

$$\gamma_j(x) = \frac{\pi_k N(x|\mu_k,\Sigma_k)}{\sum_{j=1}^{K} \pi_j N(x|\mu_j,\Sigma_j)}$$

Step 3: The parameters using the present responsibilities are re-estimated using M step as follows

$$\mu_j = \frac{\sum_{n=1}^{N} \gamma_j(x_n) x_n}{\sum_{n=1}^{N} \gamma_j(x_n)}$$

$$\Sigma_j = \frac{\sum_{n=1}^{N} \gamma_j(x_n)(x_n - \mu_j)(x_n - \mu_j)^T}{\sum_{n=1}^{N} \gamma_j(x_n)}$$

$$\pi_j = \frac{1}{N}\sum_{n=1}^{N} \gamma_j(x_n)$$

The log likelihood function is evaluated as

$$\ln p(X|\mu,\Sigma,\pi) = \sum_{n=1}^{N} \ln\left\{\sum_{k=1}^{K} \pi_k N(x_n|\mu_k,\Sigma_k)\right\}$$

If there is convergence, then the process is terminated and if there is no convergence then the procedure is returned to step 2. Table 2 shows the MSE Estimation for various architectures in Radial Basis Function Networks

Table 2 MSE Estimation for Various Architectures in RBF Networks

Architecture	Train MSE Index	Test MSE Index
1-16-1	4E-12	0.0001
2-8-2	9.47118E-05	1.50987E-07
4-4-4	0	0.001873
8-2-8	1.6E-11	1.67205E-05
28-1-1	0.000676636	5.53126E-08

4 Results and Discussion

For the in-depth Performance measures of GMM and RBF Classifier for Classification of Breast Cancer, the bench mark parameters taken for the analysis is False Alarm measures, Perfect Classification Measures, Missed Classification Measures, sensitivity and specificity measures and accuracy measures are given by the following mathematical formulae as

$$PI = \left(\frac{PC - MC - FA}{PC}\right) \times 100$$

PC is abbreviated as Perfect Classification, MC is abbreviated as Missed Classification and FA is abbreviated as False Alarm. The Sensitivity, Specificity and Accuracy measures are mathematically expressed by the following formulae

$$Sensitivity = \frac{PC}{PC + FA} \times 100$$

$$Specificity = \frac{PC}{PC + MC} \times 100$$

$$Accuracy = \frac{Sensitivity + Specificity}{2} \times 100$$

Table 3 depicts the Performance Analysis of GMM Classifier for the different stages of Breast Cancer. Table 4 shows the Consolidated Analysis of Clinical Values with the GMM and RBF neural network for the various stages Classification of breast cancer.

Table 3 Performance Analysis of GMM Classifier for the different stages of Breast Cancer

	PC	MC	FA	PI	Sensitivity	Specificity	Average
T1	100	0	0	100	100	100	100
T2	83.33	0	16.67	79.9952	83.33	100	91.665
T3	63.15	0	36.85	41.64687	63.15	100	81.575
T4	70.37	29.63	0	57.89399	100	70.37	85.185
Average	79.2125	7.4075	13.38	69.88402	86.62	92.5925	89.60625

Table 4 Consolidate Analysis of Clinical Values with the GMM and RBF neural network for the various stage Classification of breast cancer

S.No	Breast Cancer Stage	Classification (%) Through Clinical Procedure	GMM Classification Results (%)	RBF Classification Results (%)
1	T1	98	100	95
2	T2	100	91.665	91
3	T3	97	81.575	92
4	T4	100	85.185	93

5 Conclusion

The application of neural networks is a great boon to all the doctors, physicians and oncologists for the breast cancer classification. The neural network approach is highly desirable as it minimizes the biasing possibilities of the observer because it gives a good result comparison with the other classification techniques and the results are quite reliable. In this paper, the classification of breast cancer was done with three techniques such as Chi square test, GMM Classifier and RBF Classifier. The best results show a highest accuracy of 100% is obtained when GMM classifier is used in T1 stage, for T2 stage an accuracy of 91.665% is obtained in GMM classifier, for T3 stage and T4 stage RBF classifier performs better with an average accuracy of 92% and 93% respectively. Thus with the advent of neural networks and GMM classifier the different stages of breast cancer was classified successfully. Both the performance of the classifiers seem to be efficient and promising in one way or the other. Future works plan to implement the analysis of the different stage of breast cancer with different modifications in neural networks and post classification techniques.

References

[1] Perkin DM, Lara E "Estimates of the World wide frequency of sixteen major cancers" vol 41, pp 184-197, 1980.

[2] American Cancer Society, Cancer Facts and figures, Atlanta (GA), the society, 1996.

[3] Hermanek P, Sobin "International union Against Cancer TNM classification of malignant tumors, 4th Ed, 2nd revision" LH Editors, Berlin, Springer- Verlag; 1992.

[4] William H. Wolberg, W.Nick Street, and O.L. Mangasarian, "Machine learning techniques to diagnose breast cancer from image-processed nuclear features of fine needle aspirates," Computer applications for early detection and staging

[5] Lukasz Jelen, Thomas Fevens, and Adam Krzyzak, "Classification of Breast Cancer Malignancy Using Cytological Images of Fine Needle Aspiration Biopsies," International Journal of Applied Mathematics and Computer Science, vol. 18, no. 1, pp. 75-83, March 2008.

[6] R. NITHYA and B. SANTHI, "Comparative Study On Feature Extraction Method For Breast Cancer Classification," Journal of Theoretical and Applied Information Technology, vol. 33, pp. 220 - 226, November 2011.

[7] P. Filipczuk, M. Kowal, and A. Marciniak, "Feature selection for breast cancer malignancy classification problem," Journal of Medical Informatics & Technologies , vol. 15, pp. 193-199, 2010.

[8] William N. Street, "Xcyt: A system for remote cytological diagnosis and prognosis of breast cancer," Soft Computing Techniques in Breast Cancer Prognosis and Diagnosis, pp. 297-322, 2000.

[9] T. S. Subashini, V. Ramalingam, and S. Palanivel, "Breast mass classification based on cytological patterns using RBFNN and SVM," *Expert Syst. Appl.*, vol. 36, no. 3, pp. 5284-5290, Apr. 2009.

[10] M. F. Akay, "Support vector machines combined with feature selection for breast cancer diagnosis," *Expert Syst. Appl.*, vol. 36, no. 2, pp. 3240-3247, Mar. 2009.
[11] D. West, P. Mangiameli, R. Rampal, and V. West, "Ensemble strategies for a medical diagnostic decision support system: A breast cancer diagnosis application," *Eur. J. Oper. Res.*, vol. 162, no. 2, pp. 532-551, Apr. 2005.
[12] W. H. Wolberg and O. L. Mangasarian, "Multisurface method of pattern separation for medical diagnosis applied to breast cytology," in *Proceedings of the National Academy of Sciences, U.S.A., Volume 87, December 1990*, pp. 9193-9196.
[13] Y. U. Ryu, R. Chandrasekaran, and V. S. Jacob, "Breast cancer prediction using the isotonic separation technique," *Eur. J. Oper. Res.*, vol. 181, no. 2, pp. 842854, Sep. 2007.
[14] S. Sahan, K. Polat, H. Kodaz, and S. Gune§, "A new hybrid method based on fuzzy-artificial immune system and k-nn algorithm for breast cancer diagnosis.," *Comput. Biol. Med.*, vol. 37, no. 3, pp. 415-23, Mar. 2007.
[15] E. D. Ubeyli, "Implementing automated diagnostic systems for breast cancer detection," *Expert Syst. Appl.*, vol. 33, no. 4, pp. 1054-1062, Nov. 2007.
[16] Hwang et al., "Recognition of Unconstrained Handwritten Numerals by A Radial Basis Function Network Classifier,"_Pattem Recognition Letters, vol 18, pp-657- 664, 1997
[17] Sunil Kumar Prabhakar, Harikumar Rajaguru, "GMM Better than SRC for Classifying Epilepsy Risk Levels from EEG Signals", Proceedings of the International Conference on Intelligent Informatics and BioMedical Sciences (ICIIBMS), November 28-30, Okinawa, Japan

Heart Sounds Features Usage for Classification of Ventricular Septal Defect Size in Children

Kamran Hassani[1,*], Kamal Jafarian[1] and D. John Doyle[2]

[1]Department of Biomechanics, Faculty of Biomedical Engineering, Science and Research Branch, Islamic Azad University, Golzare.1, Adl Blvd, Poonak, Tehran, Iran
{K.Hasani}@srbiau.ac.ir
[2] Chief of General Anesthesiology, Anesthesiology Institute, Cleveland Clinic Abu Dhabi, Abu Dhabi, UAE
djdoyle@hotmail.com

Abstract. Ventricle septal defects (VSDs) are an important form of congenital heart disease. This study presents a new approach to VSD size estimation based on the discrete wavelet transform and artificial neural network classification of heart sounds. Heart sounds was recorded for 20 children with a VSD aged 19 ± 12 months when visiting the pediatric heart clinic of Shaheed Modarres Hospital in Tehran. The detection system was trained using 70 percent of the data and evaluated using the remaining 30%. It was found to be 96.6 percent accurate for small-size VSD ($d_{hole}<0.3d_{aorta}$) and 93.3 percent accurate for large-size VSD ($d_{hole}>0.7d_{aorta}$). Our results suggest that this approach may offer clinical utility in detecting and classifying VSDs in children.

Keywords: Biomedical signal processing, Discrete wavelet transforms, Neural networks, Phonocardiography, VSD.

1 Introduction

Ventricular septal defect (VSD), a "hole in the heart", is a common congenital heart defect presenting at birth [1]. VSD occurs as a hole in the separating wall of the heart's lower septum and allows blood to shunt from left to right side of the heart. The oxygen-rich blood then gets pumped back to the lungs instead of out to the body, causing the heart to work harder. VSDs are divided into three groups by the hole size: small- (defect diameter less than 5 mm), medium-, and large-sized VSDs. When the VSD size is large, the supplementary blood is pumped into the pulmonary artery and for this reason, lungs must work harder and they may become congested. If the defect is small, in most cases these problems do not occur, then surgery and other treatments are not required often as the defects close on their own [2]. The disease is generally diagnosed by chest X-ray, electrocardiography, ECG and auscultation [3]. Among all these methods, auscultation (listening to heart sounds) using a stethoscope is the simplest and most common diagnostic method. Auscultation is widely used in evaluating heart function during medical checkups, assisting in VSD detection [4], fetal heart beat analysis [5], diagnosing valvular heart diseases and diagnosing heart failure [6]. Heart sounds are produced by vibrations induced by valvular closure, uncommon valvular widening, chest vibrations, and abnormal blood flows [7].

In the frequency domain, many researchers including Syed et al. [8] and Wu et al. [9], struggled to take out the characteristics of local frequency analysis. To classify cardiac sounds, artificial neural networks (ANN) are a good-quality method regularly employed as a classifier. Using ANN and discrete wavelet transform based features, this paper identifies two types of VSD. As observed in this section, there was an introduction into the types of VSD and common diagnostic methods. Second part introduces method. The stages including how to provide signals, pre-processing, wavelet transform, features extraction, finally classifying signals are then explained. A review on results is presented at the third part. And the final part is allocated to discussion and conclusions.

2 Methods

2.1 Data Collection

Heart sounds were recorded for 90 seconds in 20 children with VSDs aged 19 ± 12 months at the pediatric heart clinic of Shaheed Modarres Hospital in Tehran, Iran. 10 of 20 samples were small VSDs and 10 were large. The information about the size of VSD was obtained by the echocardiographic report and by consultation with a pediatrician. Signals were recorded using a computer-based phonocardiogram (PCG) record system invented in the Sciences and Research Branch, Islamic

J. Goh et al. (eds.), *The 16th International Conference on Biomedical Engineering*,
IFMBE Proceedings 61,
DOI: 10.1007/978-981-10-4220-1_6

Azad University, Tehran, Iran [10]. This system is able to record concurrently the heart sound and the single lead ECG by a standard 44.1kHz sampling frequency with 16 bits resolution. The best place for identifying VSDs via auscultation is between third and fourth left intercostal space, so the stethoscope was inserted in this location. All experiments was conducted in an acoustic noise-isolated room.

2.2 Pre-processing

Before analysing each sample's signals, the signal was divided into 10-second pieces. Each of these HS signals was normalized by the following common relation:

$$x_{i,norm} = \frac{x_i}{\sum_{i=1}^{N} x_i^2} \quad (1)$$

In this equation, N is the number of data points, x is a HS signal, and x_{norm} is the normalized HS signal.

2.3 Feature Extraction

It is clear that the derived features play an important role in the ultimate system accuracy. Regarding to [11], we used DWT to extract 8 levels of detail coefficients. Then using 5 last levels. We used 6 different mother wavelet functions like Haar, db2, db4, db10, sym8 and coif4 shown their capabilities in previous researches. Figure 1 provides a sample recorded Heart Sound signal and obtained details using

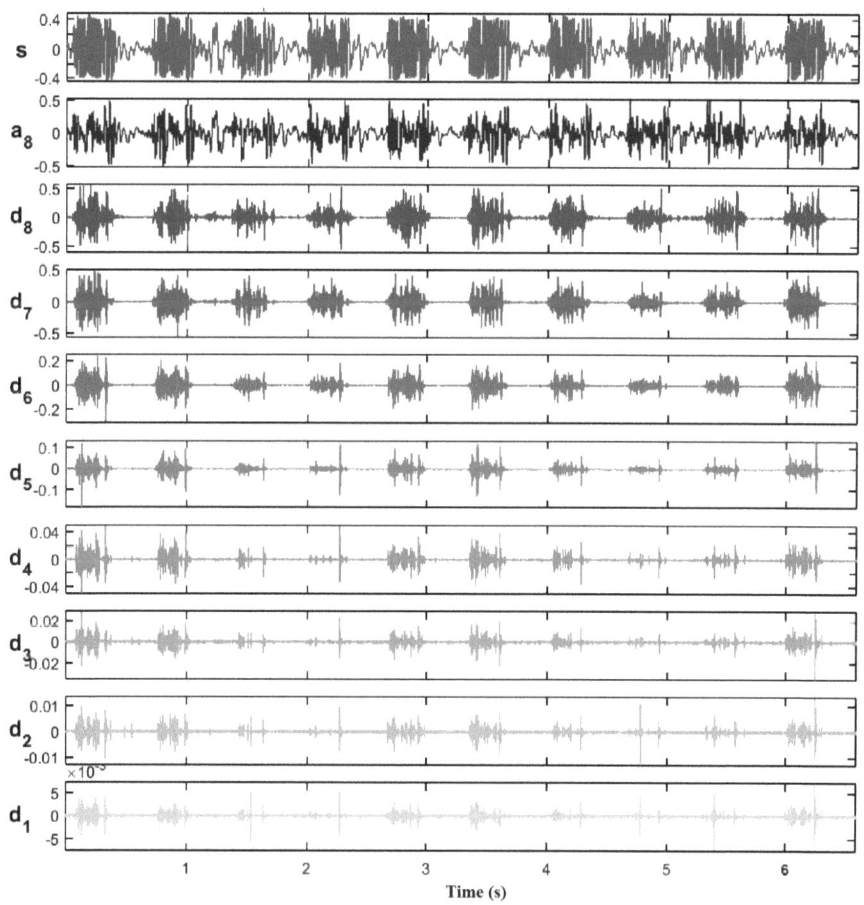

Fig. 1. Obtained details and approximate decomposition using discrete wavelet transform for a HS signal

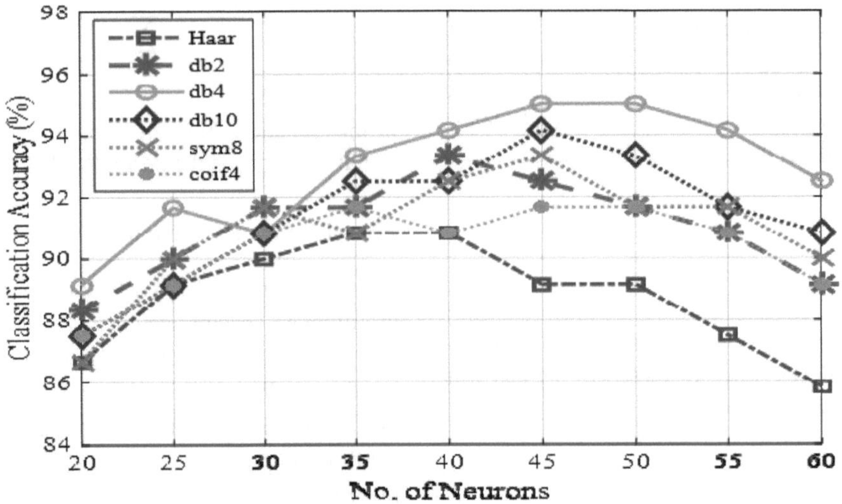

Fig. 2. Classification results for different mother wavelet types and different MLP hidden layer size

db10 function. Then we calculated power, standard deviation, Skewness, kurtosis, and Shannon entropy for each detail as extracted features. Finally, we have 25 features (5 level of details, each has 5 features), and using these features we can use a classifier to categorize VSDs among their size.

2.4 Classification

As this paper aims to find whether the septal defect is large or small among children with VSDs, in the last phase we need a perfect classifier like artificial neural network. Here, the small and large VSDs are respectively shown by 0 and 1. The artificial neural network simulates the activities of humans' nervous systems. It is, thus, considered a valuable tool for modeling complicated relations between inputs and outputs. In the state of ANN for classification, it is clear that the system output should be based on the class number of each data. Here, we used a multi-layer perceptron (MLP) neural network to classify data. The number of neurons in the hidden layer(s) and activation function for each neuron are the most important parameters for each MLP. The accuracy of the neural network classifier for correctly predicting the data class depends on the number of hidden layers, the number of neurons in each layer and the transfer function used in each layer of neurons. Reaching the best system takes a long time and in most cases the best structure is selected after several stages of trial and error.

3 Results

As explained in part 2, from both kinds of VSD studied in this research, 60 signals were achieved each had a length of 10 seconds. As mentioned above, we used 6 different mother wavelet types and for each one, we obtained specific matrix includes 25 extracted features of 120 HS signals. At the next step, we used a MLP neural network which had input, output and a single hidden layer. The number of neurons in hidden layer were changed from 20 to 60 step by 5. The activation function for all neurons was set to tangent sigmoid and Levenberg-Marquardt back-propagation method was selected as training function.

In this study, we considered 80 percent of data as train, and 20% as test data. Figure 2 summarizes the obtained accuracy of the classifier for each size of hidden layer and different mother wavelet functions after using a 5-fold cross validation training. The accuracy is the percent of data correctly classified in two classes. As shown in Figure 2, db4 is the best mother wavelet function for our purpose and by using 45 or 50 neurons in the hidden layer of MLP, we classified small and large VSDs with accuracy of 95% (96.6 percent for small-size and 93.3 percent for other group).

4 Discussion and Conclusion

Using DWT-based features extracted from 5 last levels of the discrete wavelet transform, this research classified the VSD size. We focused on the children's heart sound in this study.

The primary purpose was to determine VSD size among children only by using auscultation signal. This method was found to be 96.6 percent accurate for small-size VSD and 93.3 percent accurate for large-size VSD after complete 5-fold machine learning procedure. In other words, this scheme is 95.0% accurate to predict the size of ventricular septal defect broadly. As shown in Figure 2 these results obtained using db4 mother wavelet function and using 45 or 50 neurons in hidden layer of MLP classifier. This research shows that db4 is more capable for extracting characteristics of heart sound signal. From Figure 2, we can also find that db10 with a minor difference (less than one percent) is in second place.

Proposed multi-level DWT-based method in this study achieved almost a high accuracy for classifying VSD size, and its accuracy is not only greater than approximate accuracy of cardiologists when using auscultation, but also is comparable with the findings of other researches on diagnosing cardiac abnormalities using HS. For example, in [12], the wavelet transform was used to extract features and the neural network for classification. The accuracy of normal cardiac signals and diagnosis were 70.5 and 64.7 respectively. In [13], phonospectogram were employed to analyze cardiac murmur. The levels of specificity and sensitivity were respectively estimated at 90% and 91%. In [14], using support vector system to categorize cardiac abnormalities, Choi et al. reached the accuracy of 71 to 99.6 percent. It should be borne in mind that a direct comparison of the performance our algorithm compared to other candidate algorithms would require that they be tested against the same clinical database. For this reason, we would support the launching of a large, world-wide, internet-based phonocardiographic database that would allow investigators from around the world to collaborate and compete.

Our research results show that this approach presents favourable results in taking features out of heart sounds regarding the unique audible features of each cardiac illness, sizes of VSD in particular. This research studied the subject of determining whether the VSD size is large or small among children. By employing minor changes, this approach might also be implemented on such heart disorders as aortic stenosis (AS), pulmonary stenosis (PS), atrial septal defect (ASD), and mitral stenosis (MS) both among children (the aim of this research) or among adults.

References

1. Eisen HN. Immunology: an introduction to molecular and cellular principles of the immune response. 5th ed. New York: Harper and Row (1974)
2. Wikipedia. Ventricular Septal Defect. http://en.wikipedia.org/wiki/ventricular_septal_defect(2013)
3. American Heart Association. http://www.heart.org/HEARTORG/Conditions/CongenitalHeartDefects/AboutCongenitalHeartDefects/Ventricular-Septal-Defect-VSD_UCM_307041_Article.jsp.(2015)
4. Merck. Ventricular septal defect (VSD)). http://www.merckmanuals.com/professional/pediatrics/congenital_cardiovascular_anomalies/ventricular_septal_defect_vsd.html?qt=&sc=&alt=. (2014)
5. Cesarelli. M, Ruffo. M, Romano. M, and Bifulco. P. Simulation of foetal phonocardiographic recordings for testing of fhr extraction algorithms. Comput. Methods Prog. Biomed. (2012)
6. Kovcs, F, Horvth, C, dm Balogh, T, and Hossz, G. Fetal phonocardiography-past and future possibilities, Comput. Methods Prog. Biomed. 104 (2011)
7. Maglogiannis, I, Loukis, E, Zafiropoulos, E, and Stasis, A . Support vectors machinebased identification of heart valve diseases using heart sounds. Comput. Methods Prog. Biomed. 95 (2009)
8. Syed Z, Leeds D, Curtis D, and Guttag J. Audio-visual tools for computer-assisted diagnosis of cardiac disorders. 19th IEEE International Symposium on Computer-Based Medical Systems (2006)
9. Wu CH, Lo CW, and Wang JF. Computer-aided analysis and classification of heart sounds based on neural networks and time analysis. International conference on acoustics, speech, and signal processing, ICASSP-95, Vol. 5 (1995)
10. Sarbandi RR, Doyle JD, Navidbakhsh M, Hassani K, and Torabiyan H. A color spectrographic phonocardiography (CSP) applied to the detection and characterization of heart murmurs: preliminary results. BioMedical Engineering OnLine. 10(42) (2011)
11. Safara F, Doraisamy S, Azman A, Jantan A, and Ramaiah A.R. . Multi-level basis selection of wavelet packet decomposition tree for heart sound classification. Comput Biol Med. 43(10) (2013)
12. Andrisevic N, Ejaz K, Rios-Gutierrez F, Alba-Flores R, Nordehn G, and Burns S. Detection of heart murmurs using wavelet analysis and artificial neural networks. J Biomed Eng.127(6) (2005)
13. Noponen A, Lukkarinen S, Angerla A, and Sepponen R. Phono-spectrographic analysis of heart murmur in children. BMC Pediatr. 7(23) (2007)
14. Choi S and Jiang Z. . Cardiac sound murmurs classification with autoregressive spectral analysis and multi-support vector machine technique. Comput Biol Med. 40(1) (2010)

Multiclass Classification of APG Signals using ELM for CVD Risk Identification: A Real-Time Application

B. Niranjana Krupa[1], Kunal Bharathi[1], Manjunath Gaonkar[1], Sai Karun[1], Suhan Nath[1] and M.A.M. Ali[2]

[1]Department of Telecommunication Engineering, PES Institute of Technology, Bengaluru, India
[2] Department of Electrical Electronics and Systems Engineering, Universiti Kebangsaan Malaysia, Bangi, Malaysia

Abstract— **In this paper, we present a non-invasive method of classifying a subject's health as "Healthy" or "At Risk" of cardiovascular disease (CVD). The novelty of the work lies in recognizing the rare case of a young subject with cardiovascular disease as well as old subjects who are healthy, and the real-time implementation of CVD risk analysis. Thirty healthy and thirty pathological signals are pre-processed using Empirical Mode Decomposition (EMD), later, the analysis of the acceleration plethysmogram (APG) signals are carried out. Seven features of the wave contour are extracted along with actual age of the subject, four classes are identified using an extreme learning machine (ELM) classifier, and we made four groups which are, Healthy Young, Unhealthy Young, Healthy Old, and Unhealthy Old. Implementation of the proposed system is done on a Raspberry Pi 2 using the Python programming language. The training of the classifier and prediction of CVD risk group, using the extracted features, takes on average 17.83 milliseconds. The overall accuracy of the system is 86%.**

Keywords— **Cardiovascular diseases, APG signals, Empirical Mode Decomposition, Real-Time implementation, Extreme Learning Machine.**

I. INTRODUCTION

Cardiovascular Diseases, also called Heart Diseases, are related to the functioning of the Heart and blood vessels. Arrhythmia, Heart Attack, and Stroke are some common examples of cardiovascular disease. In the recent past, cardiovascular diseases have been responsible for approximately 17.5 million deaths every year [1].

The diminished capability of an artery to expand and contract in response to pressure changes is described by arterial stiffness whereas vessel stiffness is described by two parameters, compliance and distensibility. As the above two parameters reduce, propagation velocity of the pressure pulse along the arterial tree, called pulse wave velocity (PWV) increases. Arterial Stiffness is usually found to increase with age, though, some studies report a linear relationship with age, and others have found accelerated stiffening between the age of 50 and 60 years. Arterial stiffness, taken as a measure of carotid-femoral PWV, is an independent indicator of cardiovascular morbidity and mortality in type 2 diabetes, hypertensive patients, elderly populations, and in end-stage renal disease. Given the predictive power of PWV, discovering strategies that prevent or reduce stiffening may be significant in the prevention of cardiovascular events [2] [3].

PPG signals are recorded based on the principle of transmission or reflection of light wave of particular wavelength at the fingertips of the subject. The difference in light absorption (by the blood vessels) of oxygenated and deoxygenated blood is used find the oxygen saturation (SpO_2) by a pulse oximeter. The shape of the wave is due to the varying volume of blood at the fingertips, as the heart goes through its cycles. A number of physiological processes in the body introduce artifacts in the PPG wave, and therefore this signal can be used to monitor other processes like respiration.

There is a qualitative relation between the fingertip PPG and the pressure pulse, and therefore features of the PPG wave can be used to estimate the generalized arterial stiffness that occurs with age. However, the PPG is a relatively smooth wave and identification of inflection points is not easy [4]. Ozawa introduced two methods the first and second derivative of the PPG signal. This allows easier interpretation and more accurate recognition of the inflection points of the original PPG wave [5] [6]. It is observed that the second derivative signal which is also known as, acceleration plethysmogram (APG), is more regularly used than the first derivative. APG signals contain a lot of noise. Empirical Mode Decomposition (EMD) is used to eliminate the components that contribute to the high-frequency noise in the signal. This technique decomposes a given signal into multiple single tone functions called Intrinsic Mode Functions (IMFs). Obtaining an IMF performs as a filter on the time-series; the frequency range covered by each IMF is limited. The first IMF that always contains the highest frequencies; each following IMF contains frequencies lesser than the previous one. A zero mean test is used to identify the high-frequency components that constitute noise and not data [7].

Extreme Learning Machines (ELM) [8] are a faster implementation of Neural Network Classifier with a hidden neuron layer. The 2 features of ELM, as stated in [9], are: (a) ELM theories prove that hidden neurons are important but can be randomly generated and free from applications. (b) ELMs, which can be biologically inspired, offer major

J. Goh et al. (eds.), *The 16th International Conference on Biomedical Engineering*,
IFMBE Proceedings 61,
DOI: 10.1007/978-981-10-4220-1_7

advantages such as ease of implementation, minimal human intervention and fast learning speed. The NTU ELM website [10] provides detailed information, links and source codes for understanding Extreme Learning Machines.

The Raspberry Pi 2 is a small, low-power computer that has a 900MHz quad-core ARM Cortex-A7 CPU. The entire project is implemented using Python. The proposed work is carried out in several stages; first one is to extract various features from APG signal and second is to perform multi-level classification. In order to make these tasks feasible noise has to be removed, which is done using EMD [7] and T-test. Several cardiac risk related features such as b/a, c/a, d/a, e/a etc. are extracted using peak and valley detection algorithms. Along with these features, age of the individual, a-a interval and Augmentation index are also considered. The database comprised of 60 samples (30 healthy & 30 pathology) each of them with all the features mentioned above. These features are fed to the Extreme Learning Machine (ELM) [8] to group a sample as healthy or pathology. Ease of implementation, minimal human intervention and fast learning speed are features of ELM that makes it more suitable for real-time implementation. Using the actual age we further classify an individual as young-healthy, young-pathology, old-healthy, old-pathology. The proposed work uses optimized algorithms for real time implementation on portable platforms such as Raspberry Pi also it helps to detect abnormal cardiovascular conditions such a young individual having a pathology cardiovascular condition and an old individual having a healthy cardiovascular condition.

II. Methodology

This section provides a detailed explanation of signal preprocessing, feature extraction and classification. Fig. 1 illustrates the flow diagram of the work that is carried out.

A. Dataset overview

Data samples used in testing this work has been provided by Universiti Kebangsaan Malaysia, Biomedical Lab. A total of thirty healthy and twenty nine pathological samples recorded using Dolphin OEM 601 setup from the left hand fingertip were used. Informed consent was obtained according to Helsinki's declaration from all the subjects before recording. The signals acquired were of one and a half minute duration and the sampling frequency was 275 Hz.

B. Pre-processing

The quality of the PPG signal depends on the properties and the location of the subject's skin at measurement, including the blood oxygen saturation, blood flow rate, the individual skin structure, the measuring environment and skin temperatures.

These factors generate several types of additive artifacts which may be confined within the PPG signals. This may affect the extraction of features and hence the overall diagnosis, particularly, when the PPG signal and its derivatives are assessed in an algorithmic fashion [5]. Power line interface, DC shift, Motion artifacts (baseline wandering) and Low amplitude PPG signal are some of the major challenges in processing the PPG signals. To remove DC shift in PPG signal Linear detrend [11] is applied. Linear detrend is a curve fitting method of signal detrend where the least square fit line is subtracted from the data set. Methods based on the Fourier transform are nearly synonymous with frequency domain processing of signals. There is no doubt about how incredibly powerful Fourier analysis can be. On the other hand, its effectiveness and popularity have a downside. From the Fourier frequencies, in the context of Fourier methods, are a collection of the individual frequencies of periodic signals that a given signal is composed of. This does not mean that Hilbert spectral analysis can replace Fourier, but that it provides an alternative understanding of frequency, and an alternative view of nonlinear and non-stationary phenomena. Since PPG (also VPG) signals are non-stationary and nonlinear in nature, the conventional wavelet and Fourier based approach leads to poor results. So to remove high-frequency noises and baseline wandering EMD is used. In EMD PPG signal is decomposed into several monotone signals called IMFs using sifting process. Decomposition is stopped when standard deviation (equation 1) between current (h_k) and previous IMF (h_{k-1}) is less than a threshold (usually 0.3).

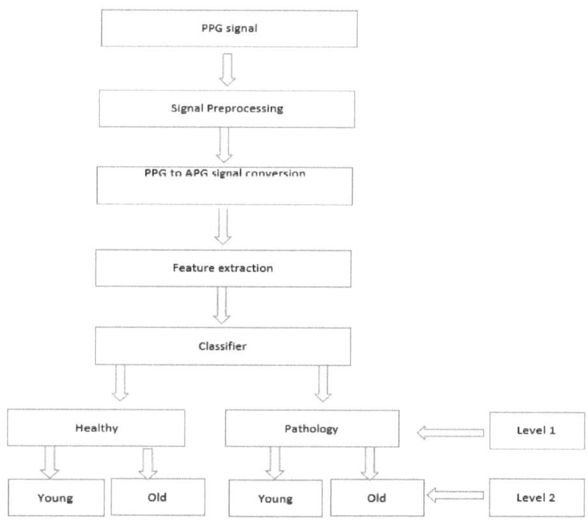

Fig. 1 Flow diagram

$$SD = \sum_{t=0}^{T} \frac{|h_{k-1}(t) - h_k(t)|^2}{h_k^2(t)} \quad (1)$$

Based on our knowledge about high frequency noise, we can safely conclude that the mean of the noise components will be close to 0. Therefore, in order to determine which IMF components contain noise in them, we need to identify those components whose mean is 0, as in (2) . This is called the Null hypothesis. We are initially going to assume that a given component has 0 mean. The T-test [12] can be used to now verify this hypothesis. If the T-test rejects our hypothesis, then the given sample does not have 0 mean, as in (3), and therefore is not a noise component.

Mathematically we can sate:

$$H_0 : \mu_{IMF} = 0 \quad (2)$$

$$H_a : \mu_{IMF} \neq 0 \quad (3)$$

A significance level of 5% is used. The probabilities are calculated from T-distribution table. There is one issue with the above method. If we perform the T-test on the IMF components, beginning from the highest frequency to the lowest frequency, we are in danger of losing data components that may have a zero mean. This problem is overcome with the help of partial sums. While using partial sums, the current IMF component, along with all those components whose frequencies are higher than this component, are added together and then fed to the T-test. The moment the system detects the highest data component, i.e. T-test rejects the null hypothesis for the first time, stop this process and reconstruct the signal from the remaining components. Out of sixty signals one pathology signal has failed to pass EMD and t-test, hence finally only twenty nine pathology signals are used.

C. Signal conversion and segmentation

The reconstructed signal is passed through FIR bandpass filter twice to get APG signal. From power spectral density of PPG signal it is evident that major frequency band of PPG lies between 0 and 20 Hz. So, an FIR filter of order 50 is used. Since amplitude swing of an Input signal will not remain constant throughout the recording, to arrive at the correct feature values Input (APG) signal is divided into 3 overlapping segments and then for each window features are extracted. Final features are mean values of the features over these 3 segments. The advantage of this method is the false rejection of peaks in peak detection can be reduced

D. Feature extraction

As shown in Fig. 2, several vascular age-related features can be derived by computing amplitude of APG wave at different points.

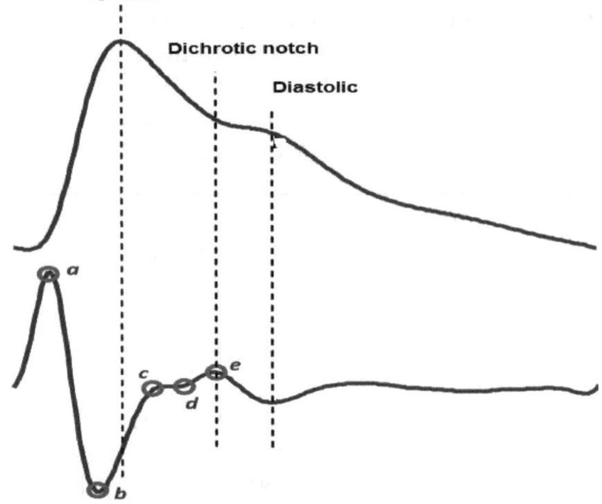

Fig. 2 Photoplethysmogram and Acceleration Plethysmogram Wave

a) Ratio b/a:

It is known to increase with age and indicates increased arterial stiffness [6]. According to a proof provided in [9] the magnitude of *b/a* of the APG is associated with the distensibility of the peripheral artery.

b) Ratio c/a:

This ratio indicates decreased arterial stiffness [6], hence it lessens with age. In, addition, it was also discovered this ratio can differentiates subjects with essential hypertension from the healthy [5] [13].

c) Ratio d/a:

It indicates decreased arterial stiffness [6], and hence decreases with age. Furthermore, it is helpful in the evaluation of vasoactive agents, and is also an index of left ventricular afterload [5].

d) Ratio e/a:

An increase in the ratio indicates a decrease in arterial stiffness [6], and hence it lessens with age [5].

e) Ratio (b-c-d-e)/a:

It was discovered that this index increased with age [6] and could be useful in the evaluation of vascular aging and also for the inspection of arteriosclerotic disease.

f) Ratio (c+d-b)/a:

This is a more inclusive aging index proposed in [14] which increased with age.

g) a-a Interval:

The R-R interval in the ECG signal correlates strongly with this interval in the APG signal as they together represent a complete heart cycle.

h) Augmentation Index (AI):

This index measures how much the reflected wave contributes to the systolic arterial pressure, and it is obtained by measuring the reflected wave coming from the periphery to the center. It has been determined as the ratio of the amplitude of late systolic peak to the early systolic peak of PPG signal.

Feature extraction follows the following algorithm:

- Segment the wave into three overlapping windows.
- For each window perform 'a' wave detection.
- Divide each window into multiple cycles of APG.
- Perform Peak and Valley detection.
- Sort all peaks, map index to amplitude.
- Reject falsely detected peaks and valleys.
- Calculate the above mentioned features.

All computed features are tabulated and targets are assigned to these signals as '1' and '0'. Where '1' represent the healthy sample and '0' represents the pathology sample. Also, measures have been taken to overcome the real-time implementation challenges such as missing peaks, valleys and run time errors.

E. Classification

To improve the efficiency, two level classification approach is used. In first level ELM is used to classify a sample as healthy or pathology, in second level simple conditional statement is used to further classify samples as young-healthy, young-pathology, old-healthy, old-pathology. ELM classifier uses 15 hidden neurons and sigmoidal function as activation function. Confusion matrix is obtained, sensitivity and specificity are calculated.

The ELM for SLFNs indicates that hidden nodes can be randomly generated. The mapping of the input data to L-dimensional ELM random feature space, and the network output is as in (4)

$$F_L(x) = \sum_{i=1}^{L} \beta_i h_i(x) = h(x)\beta \quad (4)$$

β = [β1, ..., βL]' represents the output weight matrix between the hidden nodes and the output nodes. For the better accuracy, K-fold cross validation is used and average efficiency is computed.

III. Results and Discussion

The detailed performance of the feature extraction and classification is discussed in this section. Also, the role of preprocessing is indicated by presenting the necessary plots. DC shift in the input PPG signal shown is Fig. 3(a) has been removed then EMD is applied. Based on the T-test result first two high frequency IMFs and last IMF is rejected. The reconstructed signal is passed through the filter and APG signal which is shown in Fig. 3(b) is obtained.

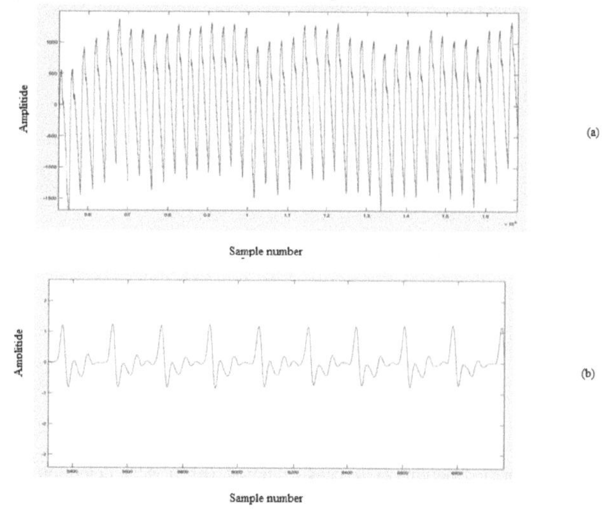

Fig. 3 PPG signal before (a) and APG signal obtained after (b) preprocessing

This APG signal is free from artifacts. A total of thirty healthy and twenty nine pathological signals are used as one pathological signal failed the t-test. Using the feature extraction algorithm mentioned earlier, 'A' wave detection in each window is carried out as shown in Fig. 4(a). 'A' peak detection is done using thresholding technique by keeping minimum peak height in the window as 70 % of the maximum peak height available. Some 'A' peaks are not getting detected because their peak height is less than 70% of the maximum peak height, by taking detected 'A' points as the reference from each window of APG signal multiple cycles of APG are obtained. Peak and valley detection on each cycle are performed as displayed in Fig. 4(b) and Fig. 4(c). In each cycle points A, B, C, D, E is detected if any of these peaks are missing, appropriate exception handling is carried out. Redundant peaks are neglected after sorting the peaks in ascending order.

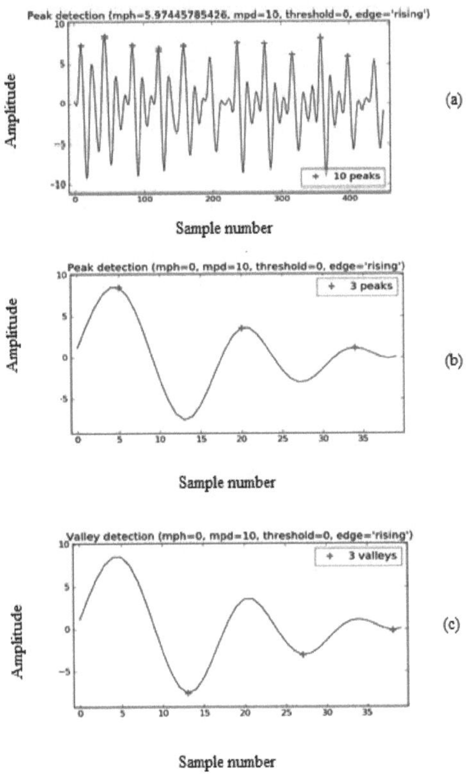

Fig. 4 Feature extraction (a) 'a' peak detection, (b) 'b' peak detection, (c) c: valley detection

Table 1 Variation of efficiency with respect to No. of initial neurons

Number of Initial Neurons 'X'	Accuracy
5	84
10	84
15	86
20	84

For the first level classification ELM classifier with sigmoidal function as activation function is used. The features as discussed in section 2.4 are given as the input to the classifier, the output of the classifier determines if the given sample belongs to healthy or pathology category. With K-4 cross-validation the average accuracy of 86% is achieved. Table 1 shows the variation of accuracy w.r.t number of hidden neurons. As we can observe from Table 1 that for accuracy is more 15 hidden neurons.

From confusion matrix shown in Table 2 we obtained a sensitivity of 91% and specificity of 80% is. Once these results are obtained from ELM classifier further classification is done, based on conditional statement using age, as young-healthy, young-pathology, old-healthy, old-pathology. This method of multi-level classification improves the accuracy of the classification.

Table 2 Confusion matrix

	Predicted: NO	Predicted: YES
Actual: NO	22	6
Actual: YES	2	25

IV. Conclusions

This study shows that it is possible to estimate the risk of cardiovascular disease using inexpensive and noninvasive equipment. Classification can be carried out in an acceptable time-frame on a portable device like the raspberry pi 2. This paper has also shown the usefulness of EMD and T-tests in APG signal denoising (and the implementation of the same on low power processors). An overall accuracy of 86% showed that the real-time application developed can be used to classify APG signals as 'Healthy Young, Unhealthy Young, Healthy Old, Unhealthy Old'. As a future work we would like to test the performance of the algorithms on a more comprehensive dataset.

Acknowledgment

The authors would like to thank the management of PES University for sponsoring this project. We would also like to thank the Biomedical Research Lab of Universiti Kebangsaan Malaysia for providing data to test the proposed work

Compliance with Ethical Standards

Ethical approval: All procedures performed in studies involving human participants were in accordance with the ethical standards of the institutional and/or national research committee and with the 1964 Helsinki declaration and its later amendments or comparable ethical standards.

Informed consent: Informed consent was obtained from all individual participants included in the study.

Conflict of Interest

The authors declare that they have no conflict of interest.

References

1. World health organization: www.who.int/mediacentre/factsheets/fs317/en/
2. Cecelja M, Chowienczyk P (2012) Role of arterial stiffness in cardiovascular disease. JRSM Cardiovasc. Dis. 1:1-11. doi: 10.1258/cvd.2012.012016
3. Laurent S, Cockcroft J, Van Bortel L et al (2006) Expert consensus document on arterial stiffness: methodological issues and clinical applications. Eur Heart J 27:2588–2605. doi:10.1093/eurheartj/ehl254
4. Alberto A (2002) The finger volume pulse and assessment of arterial properties. J Hypertens 20(12):2341-2343
5. Mohamed Elgendi (2012) On the Analysis of Fingertip Photoplethysmogram Signals. Curr. Cardiol. Rev. 8:14-25. doi: 10.2174/157340312801215782
6. Takazawa K, Tanaka N, Fujita M et al (1998) Assessment of vasoactive agents and vascular aging by the second derivative of photoplethysmogram waveform. Hypertension 32(2):365-370
7. Blanco-Velasco M, Weng B, Barner KE (2008) ECG signal denoising and baseline wander correction based on the empirical mode decomposition. Computers in Biology and Medicine 38:1-13. doi:10.1016/j.compbiomed.2007.06.003
8. Huang G-Bin, Zhu Q-Y, Siew C-K (2004) Extreme learning machine: a new learning scheme of feedforward neural networks. In: IEEE International Joint Conference on Neural Networks, vol 2, p 985, July 2004. doi:10.1109/IJCNN.2004.1380068
9. NTU Singapore: www.ntu.edu.sg/home/egbhuang/pdf/IEEE-IS-ELM.pdf
10. NTU Singapore: www.ntu.edu.sg/home/egbhuang/
11. University of Arizona: http://www.ltrr.arizona.edu/~dmeko/notes_7.pdf
12. Kothari CR (2004) Research Methodology: Methods and Techniques.
13. Nousou N, Urase S, Maniwa Y, Fujimura K, Fukui Y (2006) Classification of Acceleration Plethysmogram Using Self-Organizing Map. In: International Symposium on Intelligent Signal Processing and Communications. p 681, Dec 2006. doi: 10.1109/ISPACS.2006.364747
14. Sano, Y. Kasokudo Myakuha ni kansuru Kenkyuu no Gaiyou (in Japanese) (2003) http://jsspot.org/sano/.

Corresponding author:

Author: Prof. Dr. B. Niranjana Krupa
Institute: PES Institute of Technology
Street: 100 Feet Ring Road
City: Bengaluru
Country: India
Email: bnkrupa@pes.edu

Development of an EMG-based Human-Machine Interface on Open-source Linux Platform for Evaluating the Motor Skill Acquisition Process

Guanghao Sun[1] and Wenwei Yu[2]

[1]Graduate School of Informatics and Engineering, The University of Electro-Communications, Japan
Guanghao.Sun@uec.ac.jp
[2]Graduate School of Engineering, Chiba University, Japan
yuwill@faculty.chiba-u.jp

Abstract. Human-machine interface (HMI) not only assists disabled people to perform tasks and activities, but also helps in acquisition of motor skills for rehabilitation. Therefore, it is critically important to qualitatively assess the motor skill acquisition process during the operation of HMI. Some of the recent studies observed and analyzed that changes in bio-signals were closely related to the improvement of motor skill. This paper presents a HMI for balancing the control of virtual inverted pendulum using electromyography (EMG) signals. The proposed HMI was developed using an open-source control software RTAI (Real Time Application Interface) on Linux platform. EMG signals from the forearm were recorded using two EMG electrodes placed around the extensor carpi radialis longus and flexor carpi ulnaris muscles. Moreover, the motor skills were evaluated by classifying the EMG signal patterns (i.e., muscle activation patterns), and measuring the motion of inverted pendulum including angular position, speed, and acceleration. An experiment was conducted using the proposed HMI with 8 subjects in laboratory. The results showed that the proposed HMI is useful to examine the motor skill acquisition process, as the EMG signal patterns closely resembled to that of human motor process.

Keywords: Human-machine interface, motor skill, EMG, Open-source Linux.

1 Introduction

With the advancement of computing and bio-sensing technology, a variety of human-machine interfaces (HMIs) are being intensively studied to augment human abilities [1, 2]. Many bio-signals can be sent to a machine/computer as command signals to control it. For instance, a functional prosthetic hand was developed for improving the daily reaching and grasping activities of disabled people using surface electromyography (EMG) signal [3].

Such assistive HMI not only assists disabled people to perform tasks and activities, but also helps in the acquisition of motor skills for rehabilitation [4]. To develop a user-friendly HMI, it is necessary to make a qualitative assessment of the motor skill acquisition process during its operation. In the conventional approach, the accuracy of motion and time for completing a task is usually used for evaluating scores. However, some of recent studies indicated that observe and analyze the changes of bio-signals were closely related to the improvement of motor skill [5].

For this study, we developed an EMG-based HMI for evaluating the motor skill acquisition process via analysis of EMG signals. The proposed HMI design phases consisted of signal acquisition, signal processing, feature classification, and performance evaluation. First, we adopted an open-source signal acquisition hardware and signal processing software on a real-time Linux operating system. Next, a simple inverted pendulum simulator was constructed digitally. The EMG signals from the forearm were used for balancing control of this inverted pendulum. Then, a neural-network classifier was trained using the EMG data, and finally the motor skill acquisition process was analyzed offline and eventually evaluated in the laboratory by conducting experiments on the subjects.

2 Methods and Material

2.1 EMG-based HMI hardware development and software configuration on Linux platform

The HMI is developed on a low-cost and open-source digital control software RTAI (Real Time Application Interface, for Linux) [6]. A block diagram of the proposed HMI is shown in Fig. 1. Subject operated the HMI for balancing control of virtual inverted pendulum on a moving cart using EMG signals. The EMG signal was defined as pushing the cart to the left, during wrist flexion, whereas, pushing the cart to the right, during wrist extension. EMG signals are obtained from two EMG sensors (Sikikou Engineering Ltd., TYE-1000M, Japan) attached to the forearm around extensor carpi radialis longus muscle and flexor carpi ulnaris muscle. A data acquisition card (Incite Technology Ltd., USB-DUX-fast, UK) converted the signals to digital signals with a sampling rate of 1000 Hz. The EMG signals from the

J. Goh et al. (eds.), *The 16th International Conference on Biomedical Engineering*,
IFMBE Proceedings 61,
DOI: 10.1007/978-981-10-4220-1_8

forearm used for balancing control of the inverted pendulum simulator, was a two-dimensional animation created via OpenGL (a computer graphics library). The physical parameters and values of inverted pendulum are shown in Table 1.

Table 1. The physical parameters of inverted pendulum.

Description	Value
Mass of the cart	1.0 kg
Mass of the pendulum	0.1 kg
Length of the pendulum	1.2 m

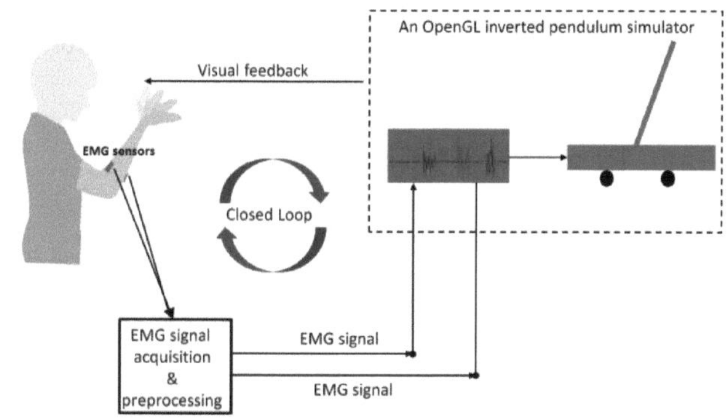

Fig. 1. EMG-based human-machine interface.

2.2 EMG Signal feature extraction and neural-network classifier for analysis of motor skill acquisition process

EMG signals from the forearm muscle were filtered using a 100 point moving average filter after full wave rectification (Fig. 2). Thereafter, time and frequency domain features were extracted i.e., percentage of EMG data to maximal voluntary contraction (MVC) and EMG power spectrum. Due to the high dimensional features of EMG, a neural network (the Kohonen's Self-Organizing Map, SOM [7]) was used for the off-line classification. SOM have been successfully used as a tool for classification in numerous studies in the past [8, 9]. The major advantage of SOM is the capability to visualize two-dimensional views of high-dimensional data as SOM maps.

2.3 Subjects and experiment procedure

Eight subjects (six males, two females) with an average age of 24 years participated in the experiment. All subjects were deemed healthy with no known history of neurological abnormalities or musculoskeletal disorders.

Each subject was seated in front of a display, and asked to control the animation of the on-screen inverted pendulum. The balancing control started with an initial setting of pendulum's angle at 0.001 radians, with increments up to 0.5 radians. The subjects continued to operate the HMI until they succeeded in gaining control over the inverted pendulum

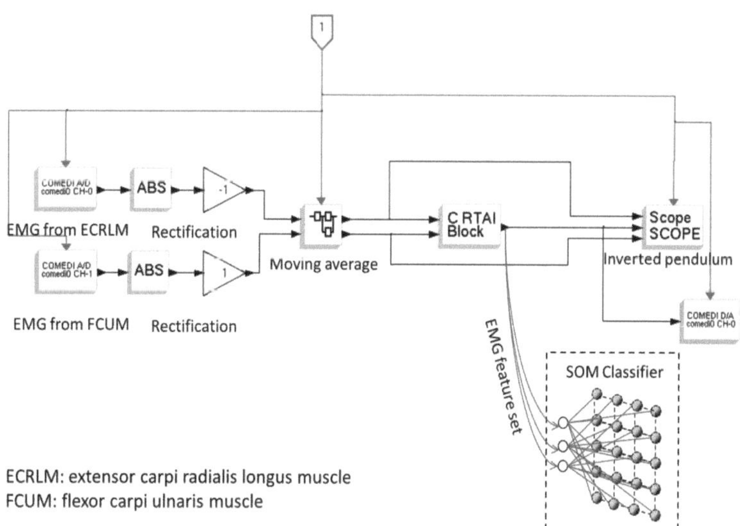

Fig. 2. Block diagram of EMG signal processing, featuring extraction, and neural network (the Kohonen's Self-Organizing Map) for classification of the EMG signal patterns.

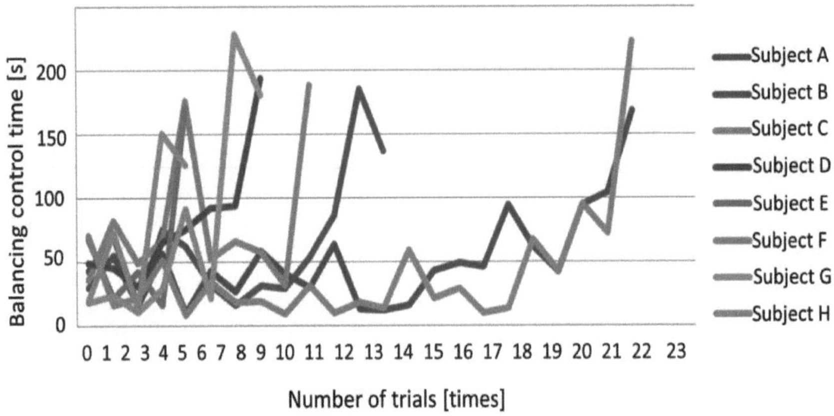

Fig. 3. Continue balancing control time and trial number for all subjects in training phase. All subjects could continuously control pendulum without failure for 180 s, when they completed the training of using HMI.

with no failure for 180 s. For the offline analysis of the motor skill acquisition process, the EMG signals and motion of the pendulum including angular position, speed, and acceleration were recorded by a laptop PC.

3 Results and Discussion

3.1 Evaluating the motion of inverted pendulum

As shown in Fig. 3, all subjects could continuously control the pendulum without failure for 180 s, after successful completion of the training to use HMI. Most subjects felt

Table 2. The standard deviation of pendulum's angle, controlled time, and %MVC in each trial

	Time [s]	Angle (SD) [radian]	%MVC	
			FCUM	**ECRLM**
Trial 1	30.1	0.18	18.0%	18.0%
Trial 2	55.6	0.15	18.0%	20.0%
Trial 3	76.0	0.11	9.0%	12.5%
Trial 4	62.6	0.18	9.0%	12.0%
Trial 5	32.4	0.12	9.0%	12.0%
Trial 6	31.9	0.14	7.5%	15.0%
Trial 7	29.6	0.18	8.5%	17.0%
Trial 8	54.5	0.12	4.5%	8.5%
Trial 9	86.3	0.1	3.5%	8.0%
Trial 10	185.6	0.1	4.0%	9.0%
Trial 11	137.1	0.11	4.0%	7.5%

FCUM: Flexor Carpi Ulnaris muscle

ECRLM: Extensor Carpi Radialis Longus muscle

SD: Standard deviation

that they achieved good control of the pendulum after repeatedly training over 8 − 13 times. Subject A and subject F showed poor performance in completion time, however continuously improved their performance.

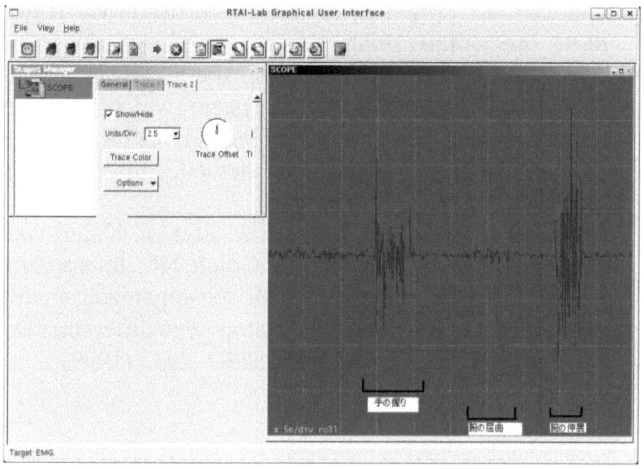

Fig. 4. The measured EMG signals from the forearm during the grasping, wrist flexion, and wrist extension motions

3.2 SOM classifier for analysis of motor skill acquisition process

The measured EMG signals from the forearm around extensor carpi radialis longus muscle and flexor carpi ulnaris muscle for subject's grasping, wrist flexion and wrist extension motions are shown in Fig. 4.

Time and frequency domain features of EMG signal (i.e., %MVC and EMG power spectrum) were input to SOM map for classification of the learning phases in each trial. The classified SOM map of the fastest learner (subject H) and the slowest learner (subject F) were compared on basis of the improvement of the motor skills. Subject H completed the training of using HMI in 5 trials, whereas subject F showed poor performance by completing the training in 22 trials. Moreover, a skill acquisition process was defined over initial, middle, and final phases. The SOM clusters corresponded to each learning phase. As shown in Fig. 5, subject H's learning phases were classified into three clusters, where the transition period from the initial to the final phase was short. However, subject F's learning phases were classified into six clusters, where the transition period from the initial to the final phase was much longer than the other subjects.

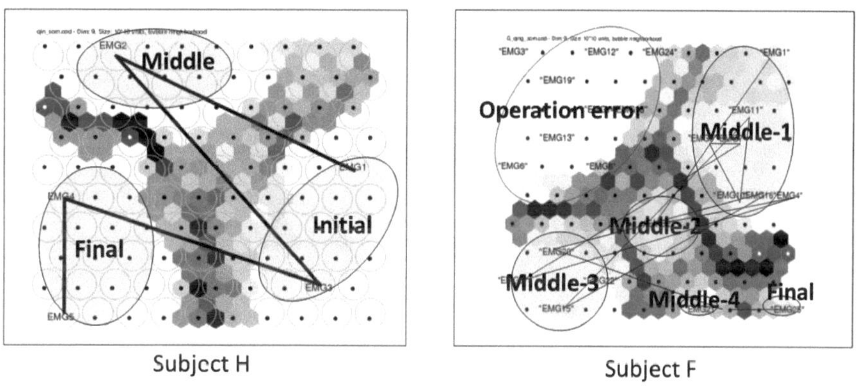

Fig. 5. SOM map for classification of the learning phases on subject H and subject F.

In order to further investigate the learning phase in each trial, the standard deviation of the pendulum's angle, controlled time, and %MVC of subject B was summarized (see Table 2). It was observed that after completing the learning phase, subject B could continue controlling pendulum without failure for 185.6 seconds. The %MVC and the standard deviation of pendulum's angle were decreased from 18.0% to 4.0% and 0.18 radians to 0.1 radians, respectively. This confirmed that subject B improved their motor skill and gained good understanding on how to operate the HMI. The same results were also be observed for the other subjects.

4 Conclusion

In this paper, an EMG-based HMI was developed for evaluating the motor skill acquisition process. The results demonstrated that a qualitative assessment of motor skill acquisition was achieved using the proposed method, as during operation of HMI, the skillfulness from EMG signal patterns closely related to that of a human's natural motor process.

References

1. Bach-y-Rita P., W Kercel S.: Sensory substitution and the human-machine interface. Trends Cogn Sci. 7(12), 541–546 (2003)
2. Tigra W., Navarro B., Cherubini A., Gorron X., Gelis A., Fattal C., Guiraud D., Azevedo-Coste C.: A novel EMG interface for individuals with tetraplegia to pilot robot hand grasping. IEEE Trans Neural Syst Rehabil Eng. in press, (2016)
3. Gonzalez-Vargas J., Dosen S., Amsuess S., Yu W., & Farina D.: Human-Machine interface for the control of multi-function systems based on electrocutaneous menu: application to multi-grasp prosthetic hands. PLoS ONE, 10(6), e0127528. (2015)
4. Muratori L. M., Lamberg E. M., Quinn L., Duff S. V.: Applying principles of motor learning and control to upper extremity rehabilitation. J Hand Ther., 26(2), 94–103 (2013).
5. Kita K., Kato R., Yokoi H., Arai T.: Analysis of skill acquisition process: A case study of arm reaching task, Rob Auton Syst, 57(2), 167–171 (2009)
6. Bucher R., Dozio L.: CACSD with Linux RTAI and RTAI-Lab. in Real Time Linux Workshop, (2003)
7. Kohonen T.: The self-organizing map. Proc. IEEE, 78(9), 1464–1480 (1990)
8. Sun G., Matsui T., Hakozaki Y., Abe S.: An infectious disease/fever screening radar system which stratifies higher-risk patients within ten seconds using a neural network and the fuzzy grouping method, J Infect., 70(3), 230–236 (2015)
9. Tamayo P., Slonim D., Mesirov J., Zhu Q., Kitareewan S., Dmitrovsky E., Lander ES., Golub TR.: Interpreting patterns of gene expression with self-organizing maps: methods and application to hematopoietic differentiation. Proc Natl Acad Sci USA., 96(6),2907–2912 (1999)

Measuring Neural, Physiological and Behavioral Effects of Frustration

Babu Ram Naidu Ramachandran[1], Sandra Alejandra Romero Pinto[1], Jannis Born[1], Stefan Winkler[1] and Rama Ratnam[2]

[1] Advanced Digital Sciences Center, Singapore
[2] University of Illinois at Urbana-Champaign, USA

Abstract— **We conducted a visual search experiment with varying task-loads to elicit frustration. Eight participants were asked to sort postal codes in a computer simulation with varying levels of task difficulty, from low to high. We collected electroencephalography (EEG), galvanic skin response (GSR), and gaze tracking data, and subjective data from a NASA Task-Load-Index based questionnaire to assess frustration during task performance. Such studies can help with work-flow process planning.**

We found that low beta EEG had greater power in tasks with higher difficulty. Eye blink rate and blink duration were higher as task difficulty increased. Finally, subjective frustration scores increased with task difficulty. We hypothesize that frustration can be detected by monitoring power in the low beta band, and rate and blink of eye duration, although this is by no means conclusive. Future work will focus on creating tasks that can directly measure frustration while keeping task difficulty the same.

Keywords— **Electroencephalography (EEG); Eye-tracking; Galvanic skin response (GSR); Multi-modal; NASA Task Load Index (TLX).**

I. Introduction

The investigation of human interactions in the real world is an intriguing topic with varying applications spanning from the study of nature to the study of human behavior in varied settings such as at work or home. In that connection, observing how feelings (frustration, anxiety etc) influence or impact the execution of work tasks can offer some insight to develop countermeasures that enhance work performance when a decline in performance is detected. We hypothesize that user levels of frustration and mental exhaustion in relation to task difficulty are valuable indicators which can be utilized to come up with ways and means to enhance work process forms and planning [1, 2].

Frustration can occur in an individual when, during a task execution, hindrance or interruptions from task completion or goal attainment are experienced. Many possible factors have been identified to cause frustration at work, including both incidental and individual factors. Incidental factors include the severity of interruptions during task execution, and loss of time [3]; individual factors include emotions such as anxiety, attitude, state of mind and mood while performing tasks. If frustration is not controlled, it can cause further impediments to performance of work, more negative emotions and stress-related reactions.

II. Materials and Methods

The participants' feedback on the workload demands of the experiment was collected electronically via a questionnaire at the beginning and at selected intervals based on NASA Task-Load-Index self-reporting instrument [4]. Total number of correct responses (CR) and response times (RT) were collected continuously throughout the experiment (behavioral response data). Pick-A-Mood (PAM) characters were also used as stimuli to collect gender information and the negative mood of the participant right before the experiment and then, after the experiment (moods used: Neutral, Bored, Sad, Irritated and Tense) [5].

Electroencephalographic (EEG) signals, eye gaze patterns, and Galvanic Skin Response (GSR) were captured for analyzing physical and physiological responses. These signals were time-stamped for time-series alignment.

A. Apparatus

An EEG ASALab system (from Advanced Neuro Technology or ANT, Enschede, Netherlands) with a Waveguard head cap (32 Channels EEG) and with provision to capture GSR and other modalities was employed. GSR data was acquired by attaching a finger-worn sensor to the participants' non-dominant hand index and middle fingers. GSR signals were recorded synchronously with EEG. LEDALAB, a Matlab-based toolbox [6, 7], was used to analyse the GSR data in terms of the Phasic and Tonic Skin Conductance levels (unit in microSiemens) after pre-processing (downsampled to 10Hz, 4th order IIR filter, cutoff frequency 2Hz, smoothing with moving average window of 100 span and segmentation

J. Goh et al. (eds.), *The 16th International Conference on Biomedical Engineering*,
IFMBE Proceedings 61,
DOI: 10.1007/978-981-10-4220-1_9

of data using the event triggers). EEG data was analysed using EEGLAB [8]. A 30Hz eye-tracker (SMI REDn Scientific eyetracker from SensoMotoric Instruments GmbH) was used for eye-gaze capture. A Logitech webcam (Logitech C920) captured video of the participants. The experiments were conducted in a spacious room with lighting condition kept to about 80 lux, measured using a light meter (ISO-TECH ILM 1335). Sound levels were monitored using a Soundmeter (Sauter SU 130) with ambient sound levels maintained at about 60 dB SPL (Sound Pressure Level). A Cedrus StimTracker (Cedrus Corporation, San Pedro, CA) was used to collect the event triggers due to stimulus changes (light sensor), and the user key presses from a Cedrus RB730 Response Pad. The response pad had colour coded buttons (keys), and the participant was asked to indicate his/her choices by pressing the appropriate colour-coded button according to the task given. The STIMTracker simultaneously sent out corresponding event triggers to the EEG amplifier via a 25 pin parallel port connection. Thus, timing was synchronised for EEG, GSR, Stimulus presentation and user response. The exact time stamps capturing stimulus onsets and user responses were also captured in SuperLab and screen recording of the visual stimulus presented via the eyetracking software SMI Experiment Center.

B. Participants

A cohort of healthy subjects (6M/2F, 24-55 years of age, median age of 30) were selected to participate in the study. All participants had good vision, corrected or uncorrected. Two were left eye dominant, and withal, right eye dominant. For one of the participants wearing spectacles, eye-tracking data was discarded because the eye-tracker could not reliably capture the eye-gaze patterns. Written informed consent was gathered from the participants prior to the experiment [NUS IRB Ref: B-15-038].

C. Stimuli

The stimulus presentation consists of a postal code in red font, picked randomly from a list of postal codes. A look-up list to infer from was shown to the participant on the left side of the screen, with colour coded tabs and ranges of postal codes assigned to each colored tab respectively. Participants were asked to look at the given postal code shown on screen, and compare the entry with the look-up list to make the selection of the matching categorized range the given postal code belongs to.

On the top right corner of the stimulus screen, a light sensor patch was used to enable tracking of stimulus screen changes via the CEDRUS StimTracker. This was also captured by the EEG amplifier as event triggers signifying stimulus presented or user response detected upon button press onset of the RB730 Response Pad. Visual feedback to participants included the number of trials they have to answer (as they progressed in the experiment), the number of correct entries, and also a pie chart which indicated their progress (0%, 10%, ... 100%).

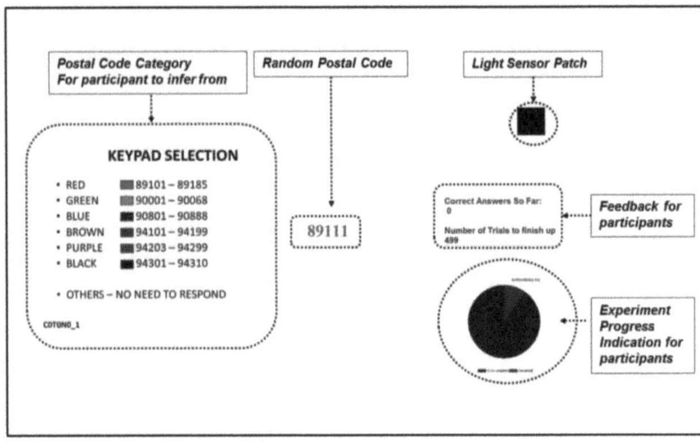

Fig. 1: Stimulus presentation (generated in Superlab 5).

D. Procedure

The stimulus presentation consisted of 9 segments in total, with the last 2 segments being identical (CTN111 and CTN111A). The tasks given to the participant varied in difficulty from low to high over 8 levels in total with 40 trials each (shown in Fig.3). Factors which contributed to task difficulty include response time, and visual search activity. Hypothetically, CTN000 was defined as the baseline segment.However, distractions were presented just before the last segment CTN111A (such as the Blue Screen of Death; 10 seconds duration) that hindered the participant from performing the task.

The arrangements of the colour codes(C), the postal code category (N) displayed on screen as well as time to respond (T) were manipulated. Brain responses (via EEG and ERP), eye gaze patterns (via Eye-tracking), physiological responses due to varying stimuli (via Galvanic Skin Response) and realtime EEG-synchronised video acquisition were collected and analysed. To keep up consistency in the test environment, the experiments were directed in low-light and relatively quiet conditions (see above).

To vary the difficulty levels, the arrangement of the colored tabs (C), the postal code range arrangement (N) and/or the time (T) given to the participant to make a choice selection after the randomly given postal code is shown were manipu-

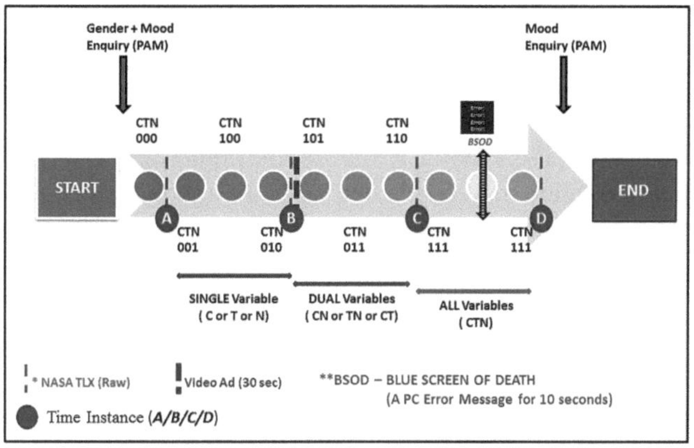

Fig. 2: Experiment protocol.

lated. The difficulty was increased sequentially, from single variable (C or T or N), dual variable (CT or CN or TN) to all variables (CTN).

Total number of correct responses (CR) and response times (RT) were collected continuously throughout the experiment (behavioral response data). The participants' feedback on the workload demands of the experiment was collected electronically at the beginning and selected intervals (time instances A/B/C/D referring to Fig. 2) based on questions adapted from the NASA Task-Load-Index (NASA TLX) self-reporting instrument, which measures the perceived demands of the experiments (mental, physical, temporal, performance, effort and frustration). The 15 piece-wise comparison was not administered in this experiment to keep the experiment under about 1 hour.

III. Results

In this paper, we report the significance of the different modalities (EEG/GSR/Eye-tracking) in correlation with the assigned hypothetical levels of difficulty in the experiment protocol. One-way Analysis of Variance (ANOVA) was done for each of the modalities with respect to the protocol and the hypothetical level of difficulty (from easy to very hard).

A. Behavioral Data

Behavioral data was collected via Superlab. In general, the rate of correct responses by the participants followed a general trend of decline from the start of the experiment (CTN000) to the end of the experiment (CTN111) as can be observed from Fig. 3. The parameters measured include the total number of correct responses, incorrect responses (also called erroneous response rate), absence of response, and the response/reaction times (RT). A correct response is recorded when the expected response (match or no response required) is the same as the participant's response. There were also instances where the participant was not required to key in any response for the given postal code, which was counted as a valid correct response for that particular trial.

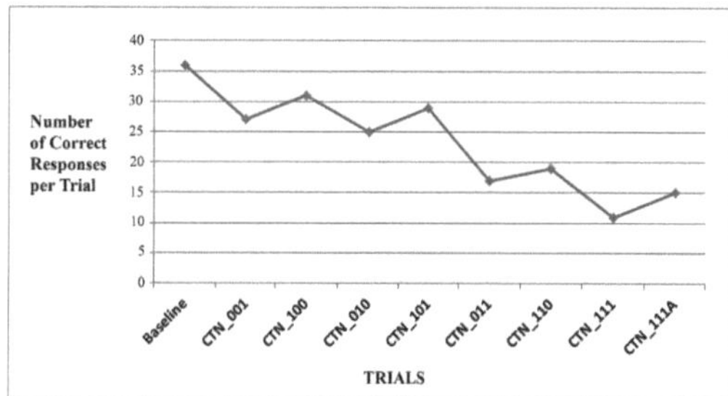

Fig. 3: Number of correct responses per segment (SuperLab).

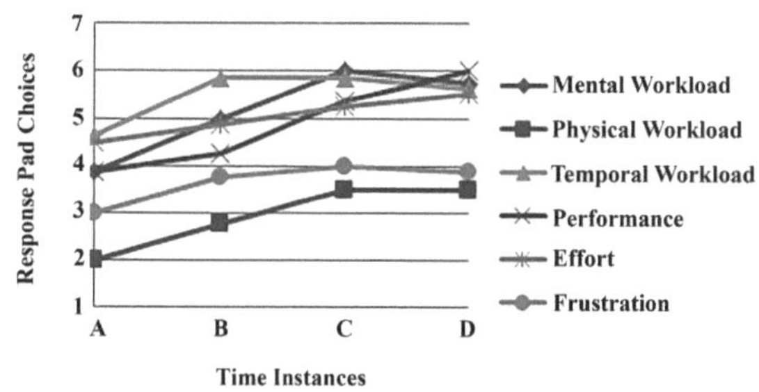

Fig. 4: Grand average NASA Task Load Index results (raw).

The GSR was elevated during CTN000, decreased afterwards and only increased again in the CTN111 and CTN111A segments. Between the CTN111 segments, the GSR was higher after the presentation of the distractors (in which the participants had to face a situation where they were unable to interact with the computer). However, in general, GSR data showed that the participants were aroused at the beginning of the experiment and during the last 2 segments of the experiment.

B. Eye Tracking Data

Blink rate increased as difficulty levels increased ($p = 0.0467$) and also in segments where time was manipulated ($p = 0.0084$). Blink duration decreased for blocks where postal code categories varied (N=1). In other words, blink durations were significantly shorter in segments where variable N was varied (condition N=1) with $p = 3 \cdot 10^{-7}$.

Fixation rate was shorter in the blocks with worst performance ($p = 0.06$). Fixation duration was longer in blocks with better performance ($p = 0.0283$).

C. EEG Data

Low beta frequency (12.5 to 18 Hz) was more prominent in later parts of the experiment when task difficulty was increasingly higher, especially in the frontal lobe, parietal and the occipital regions. Alpha frequency (7.5 to 12.5Hz) was also more prominent during the harder tasks, which we advocate was due to the task over-challenging the ability of the participants.

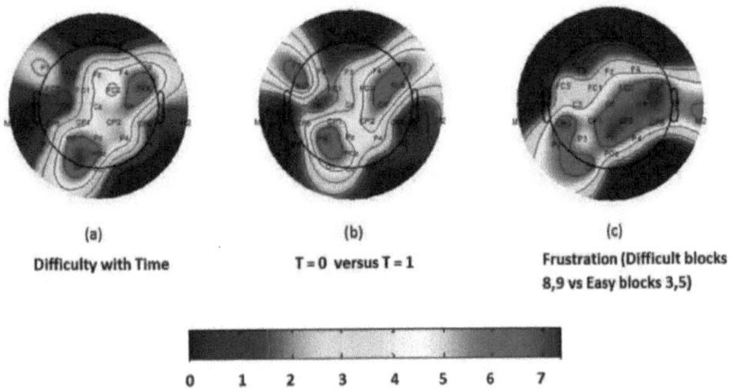

Fig. 5: EEG topography plots based on most active electrode regions.

IV. DISCUSSION

This was a nomothetic study where we studied a group of participants with a common task in similar conditions to extract multi-modal information relating to frustration. EEG, GSR and eye-tracking were used to collect the data and correlated with the participants' subjective ratings (NASA TLX based questions) and their performance (Superlab). The data was grand-averaged, and the findings indicate the elicitation of frustration in terms of Beta frequency waves (EEG), higher levels of arousal (GSR), and higher blink rates (eye-tracking) in more difficult segments. This is a preliminary study, and our population size was small. The experimental protocol demonstrated some issues. Most importantly, frustration levels were not directly controlled, but were only inferred indirectly from correlations with level of task difficulty. Further, the NASA TLX based questions should have been implemented after each segment to gather a more accurate representation of the participants subjective ratings. The 15 piecewise comparison based on the NASA TLX could yield a more accurate representation of the subjective component in this experiment. We are currently redesigning the experiment so that the level of frustration can be changed directly while maintaining a given level of task difficulty. These and other experimental design changes will be taken up in future work.

V. CONCLUSIONS

This experiment was aimed at collecting data that can provide cues and signature information about frustration onset during work tasks. While we have some preliminary evidence that frustration may be elicited based on the task difficulty, this is by no means conclusive. Future work will include modifications to the experiment protocol to directly vary frustration levels (which was not directly controlled), taking surveys at more appropriate points in the experiment, and recruiting a larger number of participants.

ACKNOWLEDGEMENTS

This study is supported by the research grant for the Human Centered Cyber-physical Systems Programme at the Advanced Digital Sciences Center from Singapore's Agency for Science, Technology and Research (A*STAR). The authors would like to thank all the participants in this pilot study.

REFERENCES

1. Klein Jonathan, Moon Youngme, Picard Rosalind W.. This computer responds to user frustration: Theory, design, and results *Interacting with Computers.* 2002;14:119–140.
2. Bessière Katie, Newhagen John E., Robinson John P., Shneiderman Ben. A model for computer frustration: The role of instrumental and dispositional factors on incident, session, and post-session frustration and mood *Computers in Human Behavior.* 2006;22:941–961.
3. Bessière Katie, Ceaparu Irina, Lazar Jonathan, Robinson John P., Shneiderman Ben. Social and psychological influences on computer user frustration in *Media Access: Social and Psychological Dimensions of New Technology Use* (Bucy Erik P., Newhagen John E.. , eds.):169–192Lawrence Erlbaum Associates 2004.
4. Hart Sandra G., Staveland Lowell E.. Development of NASA-TLX (Task Load Index): Results of empirical and theoretical research *Advances in Psychology.* 1988;52:139–183.
5. Desmet P.M.A., Vastenburg M.H., Van Bel D., Romero Herrera N.A.. Pick-A-Mood; development and application of a pictorial mood-reporting instrument in *Out of Control: Proceedings of the 8th International Conference on Design and Emotion, London, UK* 2012.
6. Benedek Mathias, Kaernbach Christian. Decomposition of skin conductance data by means of nonnegative deconvolution *Psychophysiology.* 2010;47:647-658.
7. Benedek Mathias, Kaernbach Christian. A continuous measure of phasic electrodermal activity *Journal of Neuroscience Methods.* 2010;190:80–91.
8. Delorme Arnaud, Makeig Scott. EEGLAB: An open source toolbox for analysis of single-trial EEG dynamics including independent component analysis *Journal of Neuroscience Methods.* 2004;134:9–21.

Estimating Correlation between Arterial Blood Pressure and Photoplethysmograph

Theertha Abhay, Kayalvizhi N. , Rolant Gini J.

Department of Electronics and Communication Engineering, Amrita School of Engineering, Coimbatore,
Amrita Vishwa Vidyapeetham, Amrita University, India.
n_kayalvizhi@cb.amrita.edu, j_rolantgini@cb.amrita.edu

Abstract— Photoplethysmograph (PPG) and Arterial Blood Pressure (ABP) are good indicators of cardiovascular performance. Although ABP is more widely employed, the invasive procedure of signal acquisition may cause skin rashes and inconvenience to the patients. Also, it does not allow continuous monitoring of cardiac activity. PPG, on the other hand, uses infrared light to measure the blood volume changes, which is a simple, non-invasive and can be used for continuous measurement. This paper focuses on analyzing the similarity between ABP and PPG using various features like average slope, peak position, time period, elasticity, amplitude of the signal. A segmentation algorithm was used to segment out cycles of ABP and PPG from physionet database taken from 19 patients with respiratory failure and the values of each feature were extracted for each person. Considering the population, using Pearson's correlation coefficient, the coefficient for the average slope of the PPG and peak to peak amplitude of ABP was found to be 0.55 indicating that other factors such as vessel diameter, thickness must be considered. The upstroke time period of both ABP and PPG was found to have a small difference in the range of 0.02s to 0.1s, whereas the time period of the heart cycles remained the same irrespective of the disease or healthy condition. The peak value of both ABP and PPG was found to occur with constant time difference. The elasticity with peak to peak amplitude of ABP was found to have a correlation of 0.822, and with systolic blood pressure, a correlation of 0.7622. When considered for individuals, parameters like the diastolic average slope of PPG and systolic blood pressure were found to have a good correlation coefficient ranging from 0.6 to 0.96 among other parameters which include systolic average slope, maximum and minimum slope of PPG, and the diastolic blood pressure.

Keywords— Arterial Blood Pressure (ABP); PPG; Elasticity of artery; Segmentation;

1 Introduction

Arterial Blood Pressure (ABP) is one of the important signals used for cardiovascular monitoring. ABP can be used to derive parameters like elasticity of artery and total peripheral resistance, which are strong markers of cardiovascular disease. The procedure for obtaining ABP is invasive, this makes the strong markers not easily available for regular clinical monitoring. ABP is a continuous signal with fundamental frequency as 1Hz and this pressure fluctuation is basically caused by the pumping of heart.Photoplethysmography (PPG), discovered by Hertzman, is a non-invasive, simple and inexpensive technique that is used to measure the volume changes in the blood. Infrared light emitting diode (IR LED) can be used as a source to irradiate the blood vessel. A part of light would be reflected back corresponding to the pulsatile blood flow, which is converted in to an electrical signal using photodetector. The amount signal reflected back is inversely proportional to the amount of blood flowing through the artery [1].

If heart can be considered as a pump, ECG is the electrical impulse given to the pump, ABP and PPG are the pressure and volume details of the fluid (blood) being pumped out respectively. ABP and PPG have the same source of excitation (heart) hence very good correlation is expected between the signals. A good correlation is established between the time domain and frequency domain features of radial PPG and ABP [2].Improved correlation is established between ABP and PPG, by inducing vasodilation using local heating [3]. Pulse transit time is measured from ECG and PPG, using which systolic and diastolic pressure can be estimated [4]. Establishing a good correlation between ABP and PPG will aid the continuous, non-invasive estimation of ABP.

Stationary relationship between ABP and PPG is established. When a small number of cardiac cycles are segmented, the relationship is found to be reproducible. But when the time period was greater than 20 minutes the relationship wasn't stationary [5]. A consistent intra-beat relationship between PPG and ABP is difficult to establish though there was consistency, when the mean amplitude values were averaged. Estimation of BP using PPG can also be done without the use of ECG [6]. Two different age groups were taken, below and above 25 years and they found a good correlation between both systolic and diastolic blood pressure with the diastolic time instant and the time delay between the diastolic and systolic peak in PPG.

From healthy volunteers, four parameters of PPG, such as width of the half pulse amplitude, width of two-third pulse amplitude and rising slope time and the diastolic time were chosen [7]. The volunteers were made to rest, climb and exercise for ECG, PPG and blood pressure measurement. They found that diastolic time and systolic blood pressure has good correlation.

J. Goh et al. (eds.), *The 16th International Conference on Biomedical Engineering*,
IFMBE Proceedings 61,
DOI: 10.1007/978-981-10-4220-1_10

2 Methodology

The arterial blood pressure (ABP) and Photoplethysmograph (PPG) of 19 subjects with respiratory failure , from MIMIC II database is used to analyze and estimate the correlation between them. Features of ABP and PPG are correlated by the following the steps shown in Fig. 1.

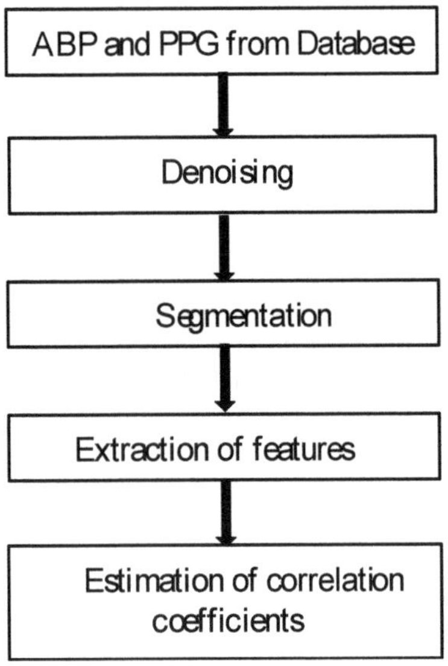

Fig. 1. Methodology Block Diagram

2.1 Filtering of ABP and PPG

Moving average filter was used to eliminate noise in all signals. With a window size (N) of 3. The larger the window length, larger the smoothing, however the sharp transitions cannot be detected for a higher window size. Moving average filter is implemented as in (1).

$$y[n] = \frac{1}{N} \Sigma_{k=1}^{N} x[n+k] \tag{1}$$

where x[n] is the input signal, y[n] is the output signal, N is the number of samples in the window and n is the sample number. All the features used in this work depend on either the peak instant or the average slope of the signal for a small period. Increasing the window size would have a serious effect on the accuracy of the feature value.

2.2 Segmentation of ABP and PPG

The filtered signals are segmented, to separate the cardiac cycles. The segmentation algorithm uses the input signal as reference to generate slope trace wave (STW). The STW generated resembles the smoothened envelope of original signal. The signal samples corresponding to one cycle of STW is segmented out as one cardiac cycle. STW is generated by following the signal during the ascending slope with corresponding sample time as T1. The condition x[n]> [n-1], is the condition of an ascending slope. When the peak value is reached the STW is retained with the peak value for

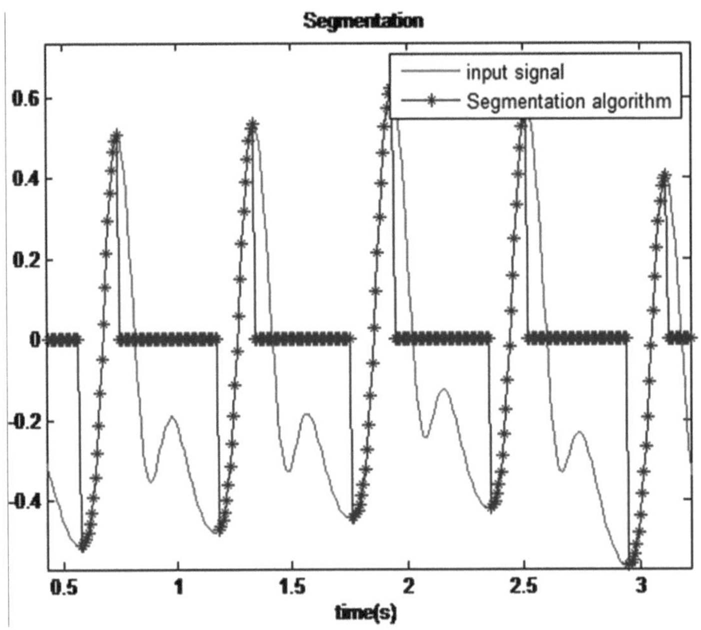

Fig. 2. Working of Segmentation algorithm

a delay period t_f. During the delay period the average slope (AS) of the input signal is calculated using (2). STW follows the input with 0.75AS as slope for a period t_f. AS is updated with the average slope of the signal during t_f. This is repeated till AS becomes a positive value. When AS becomes positive the sample time instant is recorded as T2. The input signal from T1 to T2 will be segmented as one cardiac cycle. This decrease in slope is necessary to segment the cardiac cycle with dicrotic notch. ABP or PPG can be the input signal x.

$$\text{Average slope} = \frac{\Sigma(x[i]-x[i-1])}{N} \quad (2)$$

Where,x[i] and x[i-1] sample value of the signal x with sample number i,and i-1 respectively. Fig. 2 shows the output of modified algorithm to segment out the upstroke signal.

2.3 Feature Extraction from ABP and PPG and correlation

The various features that were extracted are average slope of both ABP and PPG, time period of ABP and PPG, peak to peak amplitude of ABP and PPG, peak time instances of ABP and PPG, systolic blood pressure and diastolic blood pressure, elasticity .Elasticity was calculated by equating (3) and (4). The Bramwell-Hill equation is given by (3) and (4) is Moens-Korteweg equation . The radius, thickness of the blood vessel and the density of blood is assumed to be constant for a subject when the feature was extracted.

$$PWV = \sqrt{\frac{E_{inc}H}{2R\rho}} \quad (3)$$

$$PWV = \sqrt{\frac{dP.V}{\rho.dV}} \quad (4)$$

Where, PWV is pulse wave velocity, E_{inc} is the incremental elastic modulus, H is the vessel wall thickness, p is the density of blood, R is the vessel radius, V is the volume of the unexpended artery, that is equivalent to the average value of PPG, dP and dV represents difference in pressure and volume respectively. Correlation is used to determine how much the two sets of variables are similar.

Table 1. Different feature value of one subject for 15 cardiac cycles

cardiac cycle	Average slope of PPG	Peak to peak amplitude of ABP (mm Hg)	Time period of PPG (s)	Time period of ABP (s)	Systolic Blood Pressure (mmHg)	Diastolic Blood Pressure (mm Hg)	Elasticity (mm Hg)	Peak to peak amplitude of PPG (V)	Peak time instance of ABP (s)	Peak time instance of PPG (s)
1	4.8	37.5	0.59	0.58	115.3	77.9	4736	1.14	0.17	0.22
2	4.7	37.0	0.59	0.60	114.2	77.2	4998	1.03	0.75	0.81
3	4.5	36.8	0.60	0.60	113.7	76.9	5267	0.95	1.35	1.40
4	4.4	36.7	0.60	0.59	113.5	76.8	5140	0.95	1.95	2.00
5	4.4	36.3	0.59	0.60	113.2	76.9	5018	0.95	2.54	2.60
6	4.5	35.6	0.60	0.60	114.1	78.5	5133	0.94	3.14	3.19
7	4.6	36.9	0.59	0.58	116.2	79.3	4691	1.09	3.74	3.79
8	4.8	38.3	0.59	0.60	116.2	77.9	4563	1.21	4.33	4.38
9	4.7	38.1	0.61	0.60	115.4	77.3	5204	1.04	4.93	4.98
10	4.5	37.0	0.60	0.61	114.2	77.2	4998	1.00	5.53	5.58
11	4.6	36.5	0.60	0.60	113.4	76.9	5144	0.98	6.14	6.18
12	4.6	36.0	0.60	0.60	113.5	77.6	4932	1.00	6.74	6.78
13	4.5	36.1	0.60	0.60	115.1	79.0	4863	1.01	7.34	7.38
14	4.7	38.0	0.60	0.60	116.6	78.6	4773	1.13	7.94	7.98
15	4.5	37.8	0.60	0.60	115.4	77.6	4631	1.11	8.54	8.58

$$r = \frac{n\Sigma xy - \Sigma x \Sigma y}{\sqrt{n\Sigma x^2 - (\Sigma x)^2} \cdot \sqrt{n\Sigma y^2 - (\Sigma y)^2}} \quad (5)$$

From (5), correlation coefficient ‘r’ can be calculated. The value of r ranges between -1 to +1, where the value close to +1 or -1 is considered to have very good positive correlation or very good negative correlation. When the value is close to zer, the features are said to have poor correlation coefficient.

3 Results and discussions

For each subject the features are extracted for 15 consecutive cardiac cycle. Correlation between different features are found for 15 consecutive cardiac cycles. Table 1. shows the feature value extracted for 15 consecutive cardiac cycles of a single subject. The feature values extracted were correlated with other feature values. Table.2 shows the correlation of elasticity with peak-peak value of ABP, Systolic blood pressure and diastolic blood pressure. Not all the combination of features have good correlation. The feature combination for which the correlation is close is to one, is chosen for population study. Before doing a correlation study for a population size of 10 - 19 subjects, the feature value of a subject is taken as the average of the feature value for 15 cardiac cycles. When the average feature value is obtained for all the subjects, the feature values are normalized. Correlation coefficients for the feature combination are found. Fig.3-5 shows the correlation of features with linear fit and correlation coefficient as inset.

Table 2. CORRELATION OF ELASTICITY WITH ABP PARAMETERS

	PP-ABP	SBP	DBP
Elasticity	0.822	0.7662	-0.0172

The Table 2 shows results for a subject. From the table, we can see that systolic blood pressure has a better correlation than the diastolic blood pressure with elasticity. Fig 3 shows the correlation of normalized value of elasticity of the artery to the normalized systolic blood pressure. Elasticity value of subject calculated was not validated, because of the unavailability of the information in database. Also (3) and (4) are derived by considering the blood flow in artery as one dimensional flow, which could be one of the reason for lower correlation coefficient.

Fig 4 shows the correlation coefficient for the normalized time period of ABP and PPG. The correlation coefficient is 0.99, but there is a finite difference of 0.04s to 0.05s between peak instant of ABP and PPG. ABP peak occurred earlier than PPG peak. The finite difference can be due to effects of blood flow parameters like density of blood, varying elastic nature of the arterial wall etc., . Fig 5 shows the correlation coefficient for normalized upstroke time period of ABP and PPG. Though Fig.4 shows a correlation coefficient of 0.99 between the normalized time periods, the upstroke time period of ABP and PPG has only 0.78 as correlation coefficient. This clearly shows that pressure and volume of fluid does not vary at same rate, which can again be attributed to parameters like varying elasticity of artery.

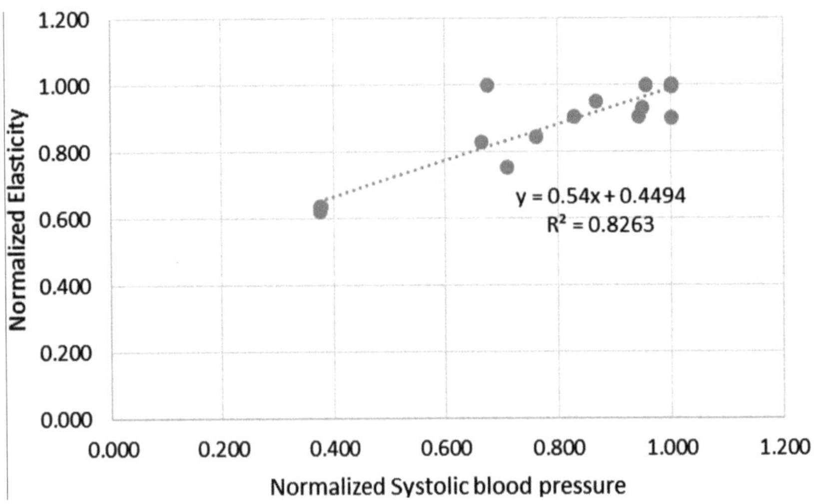

Fig. 3. Normalized Elasticity versus Normalized Systolic Blood Pressure for a population

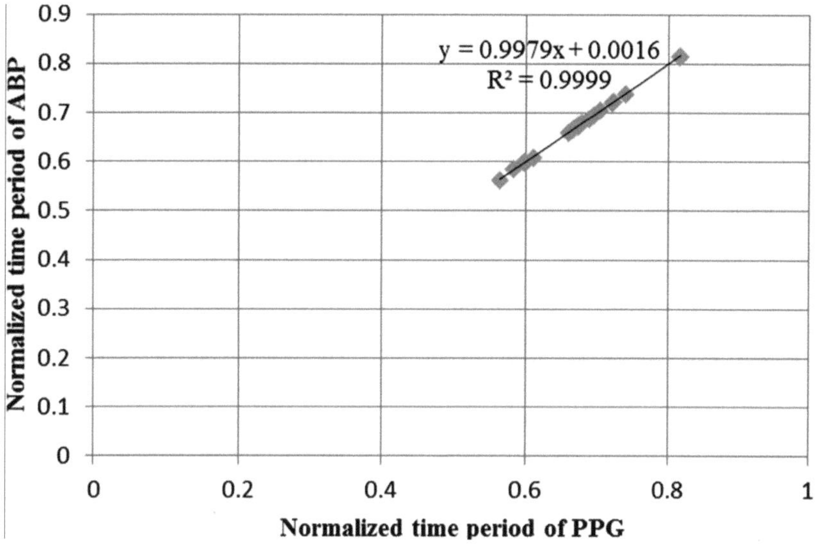

Fig. 4. Correlation between Normalized time periods of ABP and PPG.

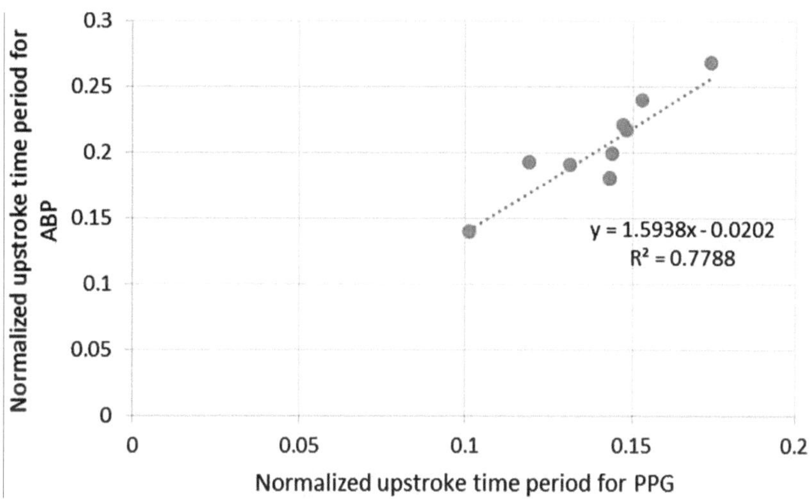

Fig. 5. Correlation of the Normalized upstroke time period of ABP and PPG

Conclusion

ABP features and PPG features of a single subject also varies between the cardiac cycles, like any other biological parameter. Hence correlation coefficient for same feature combination is not consistent across subjects. Though the signals in consideration (ABP and PPG) have a non-linear relationship, not all the feature exhibits the non-linear relationship. The time period of signals have very good linear correlation coefficient as 0.99. Heart being the source of both the signals, this result can be expected. But the other features are dominantly controlled by the other anatomical parameters rather than pumping action of heart alone, and hence showing a reduced correlation coefficient. This work can be extended with an experimental setup to validate signal analysis results.

References

1. J. Allen, "Photoplethysmography and its Application in Clinical Physiological Measurements.," *physiol. meas.,* vol. 28, pp. 1-9, 2007
2. H. Hsiu, C. L. Hsu and Tsung-Lin Wu, "A preliminary study on the correlation of frequency components between finger PPG and radial arterial BP waveforms," *2009 Internationa Conference on Biomedical and Pharmaceutical Engineering,* Singapore, 2009, pp. 1--4.

3. H. Hsiu, C. L. Hsu, C. T. Chen, W. C. Hsu, H. F. Hu, and F. C. Chen, "Correlation of Harmonic Components between the Blood Pressure and Photoplethysmography Waveforms Following Local-Heating Stimulation," *Int. J. Biosci. Biochem. Bioinforma.*, vol. 2, no. 4, 2012.
4. X. He, R. a Goubran, I. Fellow, X. P. Liu, and I. Senior, "Evaluation of the Correlation Between Blood Pressure and Pulse Transit Time," pp. 1--4, 2013.
5. P. Shaltis, a Reisner, and H. Asada, "Calibration of the photoplethysmogram to arterial blood pressure: capabilities and limitations for continuous pressure monitoring.," *Conf. Proc. IEEE Eng. Med. Biol. Soc.*, vol. 4, pp. 3970--3973, 2005.
6. R. Samria, R. Jain, A. Jha, S. Saini, S. R. Chowdhury, "Non-invasive Cuffless Estimation of Blood Pressure using Photoplethysmography without Electrocardiograph Measurement.," *IEEE*, pp. 254--257, 2014.
7. X. F.Teng, Y. T. Zhang, "Continous and Noninvasive Estimation of Arterial Blood Pressure using a Photoplethysmographic Approach.," *Proc. of 25th Int. Conf. of IEEE EMBS, Mexico*, pp. 3153--3156, 2003.
8. J. Kim, M. Kim, I. Won, S. Yang, K. Lee, and W. Huh, "A biomedical signal segmentation algorithm for event detection based on slope tracing," *Proc. 31st Annu. Int. Conf. IEEE Eng. Med. Biol. Soc. Eng. Futur. Biomed. EMBC 2009*, pp. 1889--1892, 2009
9. Goldberger AL, Amaral LAN, Glass L, Hausdorff JM, Ivanov PCh, Mark RG, Mietus JE, Moody GB, Peng CK, Stanley HE. PhysioBank, PhysioToolkit, and PhysioNet: Components of a New Research Resource for Complex Physiologic Signals. *Circulation* 101(23):e215--e220 [Circulation Electronic Pages; http://circ.ahaiournals.org/content/101/23/e215.fulll; 2000 (June 13). PMID: 10851218; doi: 10.1161/01.CIR.101.23.e215.

Flow visualization for Nasal Cavity Flow in Aerosol Exhalation Through Nose Treatment

Takahisa Yamamoto[1]*, Yoshiki Kobayashi[2], Shunpei Shikano[3], Masahiro Takeyama[3], Mikiya Asako[2], and Koichi Tomoda[3]

[1] Dept. Mech. Eng., National Institute of Technology, Gifu College, Japan
ytaka@gifu-nct.ac.jp
[2] Dept. Otolaryngology-Head and Neck Surgery, Kansai Medical University, Japan
[3] Advanced Course, National Institute of Technology, Gifu College, Japan

Abstract— Aerosol medicine exhalation through the nose (ETN) is one of promising and comprehensive treatment methods for Eosinophilic Chronic Rhinosinusitis (ECRS) with asthma. In this treatment, the patient inhales aerosol of inhaled corticosteroid (ICS) medicine from mouth using portable inhaler. Then a part of the aerosol still floats and remains in upper airway. When the patient exhales inhaled air through the nose, the aerosol is effectively transported on the walls of middle meatus and olfactory fissure. This study performed Computational Fluid Dynamics (CFD) analysis for the transport phenomena of aerosol medicine during exhalation period in order to evaluate the curative effect of ETN numerically. As a result of CFD analysis, ETN formed impinging flow toward upper wall of nasopharynx, subsequently complex swirl and circulation flow in the nasopharynx region. In addition, main flow of ETN passed upper region of nasal cavity. Such the tendencies affected on aerosol transport characteristics; a part of aerosol particles moved into ethmoindal sinuses.

Keywords— eosinophilic chronic rhinosinusitis, computer fluid dynamics, exhalation through nose

1 Introduction

Eosinophilic chronic rhinosinusitis (ECRS) is considered a refractory and intractable disease. Patients with ECRS present with asthma, thick mucus production, long-term nasal congestion, loss of sense of smell, and intermittent acute exacerbations secondary to bacterial infections. Despite medical and surgical interventions, there is a high rate of recurrence with significant impairment to quality of life. The recent increasing prevalence of ECRS in east and south Asian countries and the strong tendency of ECRS to reoccur after surgery should be considered. The majority of cases need repeat surgery, and histological examinations of these cases show eosinophilic-dominant inflammation. The degradation and accumulation of eosinophils, release of cytokines, and mucus secretion have important roles in the pathogenesis of ECRS [1]. Aerosol medicine exhalation through the nose (ETN) is one of promising and comprehensive treatment methods for ECRS [2]. In this treatment, the patient inhales aerosol of inhaled corticosteroid (ICS) medicine from mouth using portable inhaler. Then a part of the aerosol still floats and remains in upper airway. When the patient exhales inhaled air through the nose, the aerosol is effectively transported on the walls of middle meatus and olfactory fissure. Almost of all patients with ECRS obtain the efficacy of ETN. On the other hand, there are some patients who could not obtain the efficacy. The detail mechanism of how ETN improves ECRS with asthma is still controversial even though ETN gets a lot of attention as a treatment method for ECRS. In fact, it has been found that the efficacy of ETN strongly depends on expiration condition (expiration velocity, time and pattern) from clinical findings. This study performed both Computational Fluid Dynamics (CFD) analysis Experimental Fluid Dynamics (EFD) analysis for the transport phenomena of aerosol medicine during exhalation period in order to evaluate the curative effect of ETN numerically and experimentally.

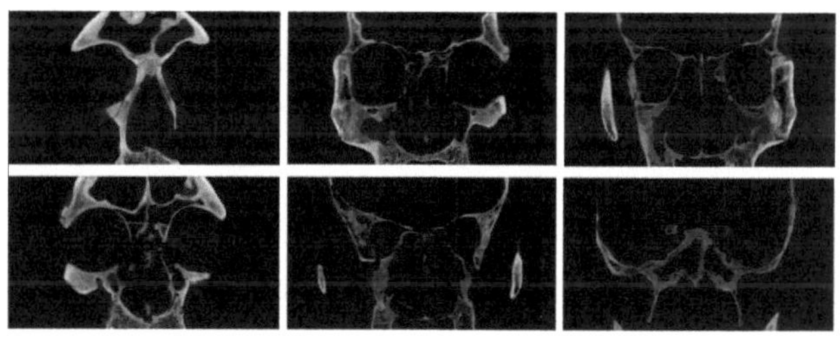

Fig. 1. 3D anatomically patient-specific model (nasopharynx)

2 Case Data and ETN Therapy

A 75-years-old male, who had ECRS with asthma and a history of endoscopic sinus surgery, was selected as an analysis case in this study. Figure 1 shows the CT images of the patient. In this case, endoscopic sinus surgery had operated and his ostium of ethomoidal sinus had been enlarged. These morphology of the patient's nasal cavity elucidate

J. Goh et al. (eds.), *The 16th International Conference on Biomedical Engineering*,
IFMBE Proceedings 61,
DOI: 10.1007/978-981-10-4220-1_11

aerosol transport phenomena in nasal-pharynx region. A 3D anatomically accurate patient-specific model was reconstructed from the data obtained using multi-detector CT scanner with medical imaging software package, Mimics (Materialise Co.) [3]. The entire series was loaded into the software, and then the nasal-pharynx airway was identified in each of the axial images based on predefined threshold of 500 Housfield units relative to the surrounding tissue. Figure 2 is 3D model of the nasal-pharynx airway. This model is composed by left and right common nasal meatuses, ethomoidal and sphenoidal sinuses, and a nasopharynx.

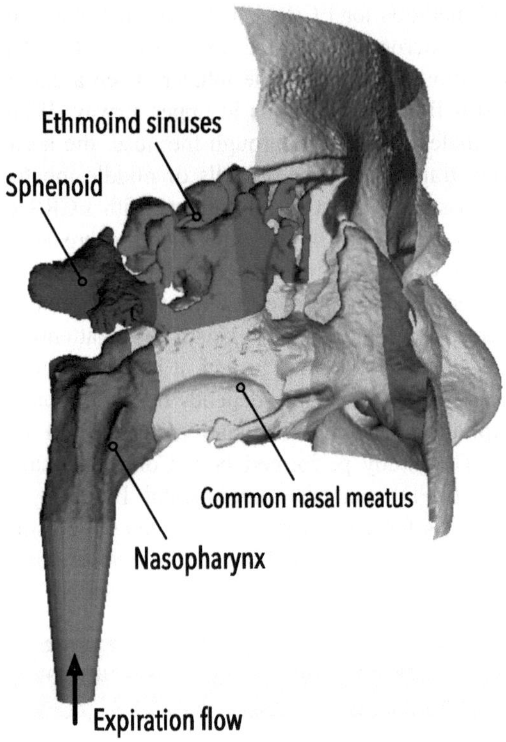

Fig. 2. 3D anatomically patient-specific model (nasopharynx)

In the ETN therapy, the patient inhales ICS aerosol medicine from mouth using portable inhaler. HFA-BDP (Quvar^{TM}) is commonly used as the portable inhaler for ETN therapy in Japan [1,2]. Inhale this medication through mouth and exhale via nose is usually two times in a day. One puff of this inhaler includes ICS aerosol medicine of 60 mg, and its mean diameter is 3 μm. According to researches reported by Yeh and Schum [4,5], the aerosol deposition on trachea branches and lung depends on diameter of aerosol and forced expiratory volume, and its deposition fraction is 50% during a inhalation period. That is half of inhaled ICS aerosol medicine reach nasal-pharynx region at the exhalation period. Transport characteristics of such the ICS aerosol medicine is significantly important to obtain the efficacy of the ETN.

3 Computer Fluid Dynamics Analysis

The nasal-pharynx airway model was exported into CFD meshing software package to generate discrete volume cells [6]. This study used both a Euler-Lagrange particle transport model for aerosol transport and a Large Eddy Simulation (LES) model for complex intranasal turbulent flow [7, 8]. These models are able to account for the transient transport of mass and turbulent energy, and consequently, provides highly accurate predictions of the amount of flow separation under adverse pressure gradients [9-11]. Expiratory flow through the nose was analyzed using the analysis models in this study. The expiratory flow inflow the nasal-pharynx airway model from the pharynx (the bottom of Fig. 2), flow through the nasopharynx and nasal cavity, and finally flow out from nostril. This study assumed forced expiratory volume 1.5 l in all analysis conditions, and considered three expiratory flow rate conditions on an inlet boundary, 30 and 90 l/min, respectively. These flow rate conditions are based on quiet breathing state of adults. The expiratory times are 3.0 and 1.0 seconds, respectively. These boundary conditions were applied in a step state at the beginning of the analysis. As for aerosol transport analysis, in order to evaluate the amount of aerosol deposition on each areas of nasal-pharynx airway, this study considered and evaluated the amount of the aerosol deposition on four areas in right and left nasal-pharynx, respectively; a nasopharynx, a sphenoid, an ethomoid sinus and a common nasal meatuses as shown in Fig. 2. The rate of inflowed aerosol medicine to nasal-pharynx region was set at 30 mg/s. These analysis conditions are summarized in Table 1.

Table 1. Summary of analysis conditions

Forced expiratory volume [l]	Aerosol inflow rate [mg/s]	Expiratory flow rate [l/min]	Expiratory time [s]
1.5	30	30	3.0
1.5	30	90	1.0

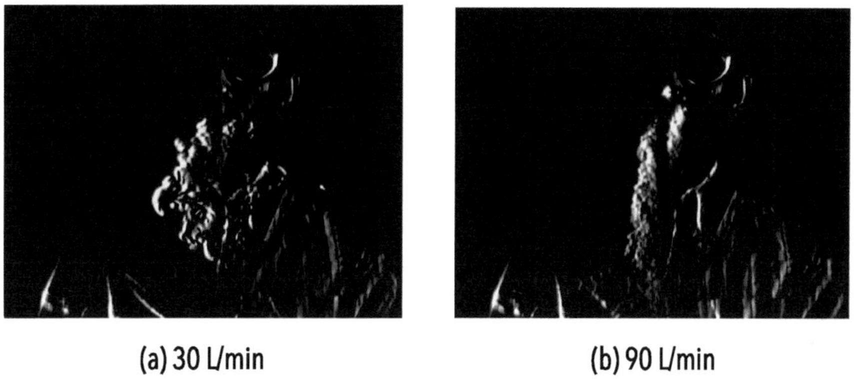

Fig. 3. Visualization of expiratory flow during exhalation through the nose: expiratory flow rate (a) 30 *l*/min and (b) 90 *l*/min

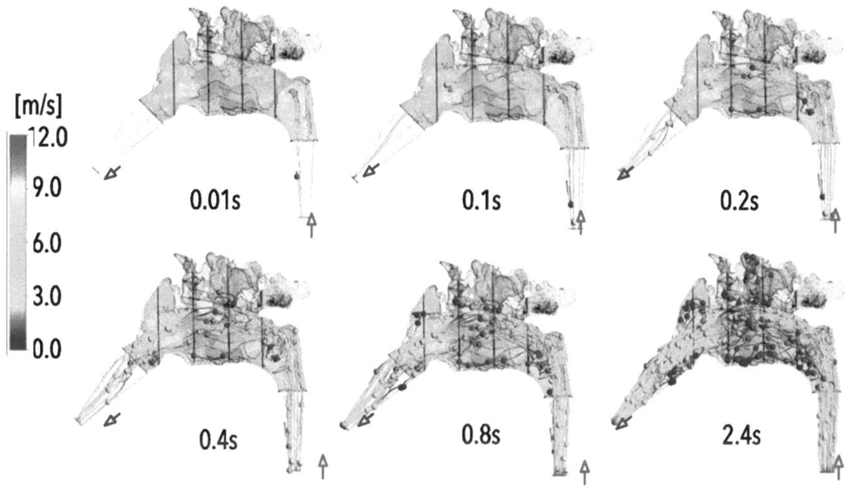

Fig. 4. Trajectories of aerosol transport in expiratory flow rate 30 *l*/min

4 Result and Disucssion

This study conducted visualization experiment of expiratory flow in the actual ETN therapy using a green sheet laser (3 W power) and a high sensitive CCD camera (particle visualization system, Shin Nippon Air Tech.). Figure 3 shows the results of the visualization experiment in expiratory flow rate 30 and 90 *l*/min, respectively. White jet flow areas are confirmed in the results. These white areas indicate that expiration flow contains ICS aerosol medicine, namely, ICS aerosol medicine can effectively reach nasal-pharynx region during exhalation period. The contrast of the white area reflects particle density of ICS aerosol medicine in the expiratory flow, and the contrasts in both conditions are almost same. The comparison of the contrasts between two conditions suggests that the particle density in the expiratory flow does not depend expiratory flow rate and and the expiratory flow rate just affects on aerosol expiratory time. Further experiment is required to make clear expiratory flow characteristics quantitatively.

Figures 4 and 5 show the results of CFD analysis, trajectories of aerosol particle during ETN. From the results, ETN formed impinging flow toward upper wall of nasopharynx, subsequently complex swirl and circulation flow in the nasopharynx region. In addition, main flow of ETN passed upper region of common nasal meatus. Such the tendencies affected on aerosol transport characteristics; a part of aerosol particles moved into ethmoindal sinuses. Table 2 shows deposition fraction of ICS aerosol medicine on each areas during ETN. The deposition depended on flow rate of exhalation. This tendency was more remarkable on the upper wall of nasopharynx. On the other hand, according with a result of transient deposition characteristics on the left ethmoidal sinuses shown in Fig. 6, the deposition rate of aerosol on the ethmoidal sinuses did not appear strong correlation with expiratory flow

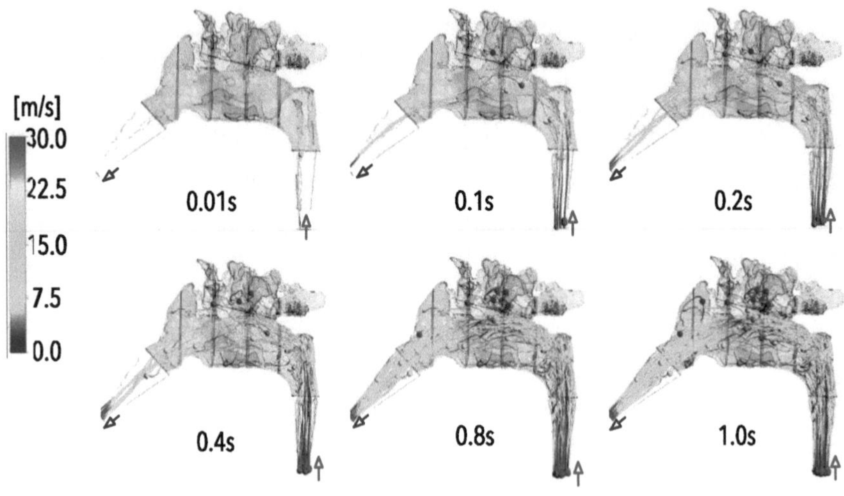

Fig. 5. Trajectories of aerosol transport in expiratory flow rate 90 *l*/min

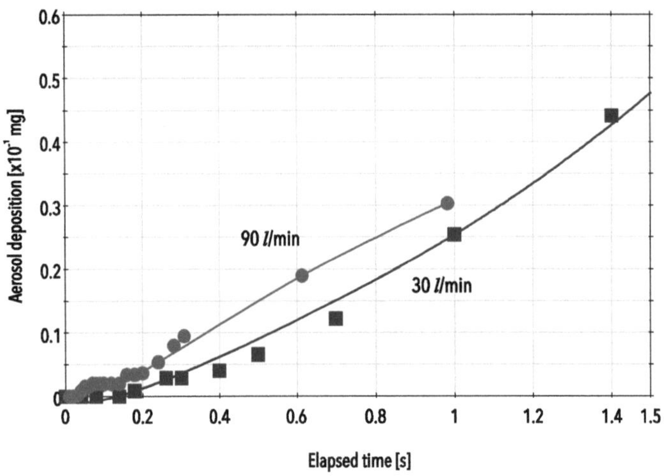

Fig. 6. Comparison of transient deposition characteristics on the left ethmoidal sinuses between expiratory flow rates, 30 and 90 *l*/min

rate. These results indicate that the phenomena of aerosol transport and deposition during ETN has non-stationary characteristics strongly. In past researches concerning CFD analysis for intranasal aerosol transport, steady-state turbulent flow model had been applied as CFD model [12-15]. New finding of this study is that unsteady turbulent model, the LES turbulent model adopted in this study, is needed in the further investigation for ETN and to reveal the mechanism how ETN treats ECRS.

Table 2. Summary of deposition fractions in nasal-pharynx region

Expiratory flow rate [*l*/min]	Common nasal meatus [%]	Ethmoid sinuses [%]	Nasopharynx [%]	Overall [%]
30	11.3	3.58	12.5	26.2
90	13.8	4.1	26.5	44.5

5 Conclusion

This study performed CFD analysis for the transport phenomena of ICS aerosol medicine during exhalation period in order to evaluate the curative effect of ETN numerically. Furthermore expiratory flow visualization was also conducted using the flow visualization system. As the result of CFD analysis, ETN formed impinging flow toward upper wall of nasopharynx, subsequently complex swirl and circulation flow in the nasopharynx region. In addition, main flow of ETN passed upper region of nasal cavity. Such the tendencies affected on aerosol transport characteristics; a part of aerosol particles moved into ethmoindal sinuses. Total aerosol deposition amount during ETN depended on flow rate of exhalation. This tendency was more remarkable on the upper wall of nasopharynx. The result of flow visualization suggests that the particle density of ICS aerosol medicine in expiratory flow during ETN therapy is not affected by expiratory flow rate and there is saturation value of the density. Further experiment is required to make clear expiratory flow characteristics quantitatively.

Acknowledgement

This work was supported by JSPS KAKENHI Grant Number No.16K06098.

References

1. Kohno, T., Yamamoto, Y., Kurokawa, M.: Effect of inhaled corticosteroid on allergic rhinitis, Asthma, 19(4), 72-76 (2006)
2. Kobayashi, Y., Asako, M., Kanda, A., Tomoda, K., Yasuba, H.: A novel therapeutic use of HFA-BDP metered-dose inhaler for asthmatic patients with rhinosinusitis: Case series. Int. J. Clin. Pharm. Th., 52, 914-919 (2014)
3. Materialize Inc.: Mimics Users Guide. Materialize Inc, Leuven, Belgium (2014)
4. Yeh, H.C., Schum, G.M.: Models of human lung airways and their application to inhaled particle deposition. Bulletin of Mathematical Biology, 42, 461-480 (1980)
5. Schum, G.M., Yeh, H.C.: Theoretical evaluation of aerosol deposition in anatomical models of mammalian lung airways. Bulletin of Mathematical Biology, 42, 1-5 (1980)
6. Lorensen, W.E., Cline, H.E.: Marching Cubes - A high resolution 3D surface construction algorithm. In: 14th annual Conf. Computer-graphics and Interactive Techniques, New York, USA (1987)
7. ANSYS Japan Inc.: ANSYS ICEM CFD Release 15.0 Manual. ANSYS Japan Inc., Tokyo, Japan (2014)
8. ANSYS Japan Inc.: ANSYS CFX-Solver Theory Guide Release 15.0, ANSYS Japan Inc., Tokyo, Japan (2014)
9. Jeong, S.J., Kim, W.S.: Numerical investigation on the flow characteristics and aerodynamic force of the upper airway of patient with obstructive sleep apnea using computational fluid dynamics. Medical Engineering & Physics, 29(6), 639-651 (2009).
10. Gemci, T., Ponyavin, V., Chen, Y., et al.: Computational model of airway in upper generations of human respiratory tract. Biomechanics, 27, 2047-2054 (2008)
11. Mylavarapu, V., Murugappan, S., Mihaescu, M., et al.: Validation of computational fluid dynamics methodology used for human upper airway flow simulations. Biomechanics, 42(10), 1553-1559 (2009)
12. Yamamoto, T., Nakata, S., Monya, M.: Computational Fluid Dynamics for intranasal heat and mass transfer. Oto-rhinolaryngology Tokyo, 52(1), 24-29 (2009)
13. Kuroda, K., Yamamoto, T., Hirose, T.: CFD analysis for medicinal aerosol transport characteristics in nasal cavity flow. In: JSME Bioengineering Conference 2015, 181. JSME, Tokyo (2015)
14. Doorly D.J., Taylor D.J., Schroter R.C.: Mechanics of airflow in the human nasal airways. Respiratory Physiology & Neurobiology, 163(1-3), 100-110 (2008)
15. Nomura T., Ushio M., Kondo K., Yamasoba T.: Effects of nasal septum perforation repair surgery on three-dimensional airflow: an evaluation using computational fluid dynamics. Eur. Arch. Oto-Rhino-Laryngology, 272(11), 3327-3333 (2015)

Multi physiological signs model to enhance accuracy of ECG peaks detection

A. Delrieu[1], M. Hoël[1], C.T. Phua[2] and G. Lissorgues[1]

[1]ESIEE-ESYCOM, Paris, France
[2]School of Engineering, Nanyang Polytechnic, Singapore

Abstract- Accurate R peaks detection in electrocardiogram (ECG) is an important process to assess the cardiovascular health of an individual (e.g. heart arrhythmia, heart rate variability, etc.). Many studies have presented various methods to detect R peaks in ECG using single physiological signal (i.e. ECG) and are highly subjective to the quality of the ECG signal. In this paper, an accurate R peaks detection algorithm is proposed based on the use of electromechanical physiological signals (i.e. ECG and photoplethysmography (PPG)). Concurrent processing of both ECG and PPG is able to reduce the need to have high quality ECG and allows the use of simple signal processing algorithms to identify the locations of R peaks in ECG signals. The flexibility of our method was demonstrated through concurrent implementation on a low cost platform (i.e. BeagleBone Black (BBB)) and FPGA platform (i.e. myRIO from National Instrument), achieving respective accuracy of 96% and 98%, using physiological signals acquired in real-time. The accuracy provided by our method is able to be applied on wearables and supports accurate real-time assessment of cardiovascular health.

Keywords- Electrocardiogram (ECG), QRS complex, R-peaks, Photoplethysmography (PPG).

I- INTRODUCTION

The electrocardiogram (ECG) is an important and well-studied biological signal used in multiple medical applications. ECG comprises of multiple features where most lifestyle application is focused on the detection of the R-peaks within the QRS complex. The ability to determine R-peak accurately is often required in R-R intervals analysis to determine ECG anomaly due to heart arrhythmia.

Many research had been conducted on methods to detect QRS complex[1-2-3]. The best known methods are the differentiation methods[4], digital filters[5-9], neural networks[10-12], filter banks[13], hidden Markov models[14], genetic algorithm[15] and maximum a posterior (MAP) estimator[16-17]. In the paper by Balda RA[18], they proposed the differentiator operator to detect QRS complex, and this method was later used by Ahlstrom and Tompkins[19], Friesen[20] and Tompkins[21] to develop methods that are able to determine the sensitivity of QRS complex to noise. Finally, methods based on the Hilbert transform[22-23-24] have the ability to distinguish between dominant peaks in signal among other peaks. Though these methods have shown good results, however they may fail in cases of low amplitude R wave.

An effective way to pre-process R-peaks generally requires the setting of a threshold value, which is manually fixed at the lowest value possible aiming at preventing the detection of P and T waves; and baseline noise. However, it is a challenge to determine a threshold value that fits the ECG signal due to baseline wondering, motion artefacts and variations in P and T waves. Xu and Li[25] have shown that using adaptive thresholding for automatic determining of threshold provides suitable results for the detection of R-peaks. However, this algorithm dependent on the quality of ECG signal acquired.

To-date, concurrent acquisition of ECG and photoplethysmography (PPG) has become possible. In this paper, an algorithm to automatically detect R-peaks in ECG with poor signal to noise ratio (SNR) through concurrent processing of ECG and PPG signals was developed. This algorithm was implemented onto low-cost embedded platforms to demonstrate accurate real-time ECG R-peaks detection.

II- METHODOLOGY

The algorithm developed in this paper is divided into 3 parts. The first part is the pre-processing of the raw ECG and PPG signals to remove environmental noise. The second part is to apply an adaptive threshold to detect dominant peaks in the pre-processed ECG and PPG signals. Finally, cross comparison of peaks detected in ECG and PPG signals is used to enhance the accuracy of R-peak detection in ECG.

A ECG and PPG signal acquisition

To acquire the ECG signal, commercially available Plessey's EPIC sensors were used. The sensors come in a pair and are used as dry electrodes held by each hand to

J. Goh et al. (eds.), *The 16th International Conference on Biomedical Engineering*,
IFMBE Proceedings 61,
DOI: 10.1007/978-981-10-4220-1_12

acquire ECG signal that is equivalent to Lead II in the traditional wet electrode ECG signal. The setup and typical waveform acquired is shown in Figure 1.

To acquire the PPG signal, commercially available Easy Pulse Sensor HRM-2511e was used and the waveform acquired is illustrated in Figure 2.

The ECG and PPG signals are concurrently digitised using the BeagleBone Black and myRIO from National Instrument.

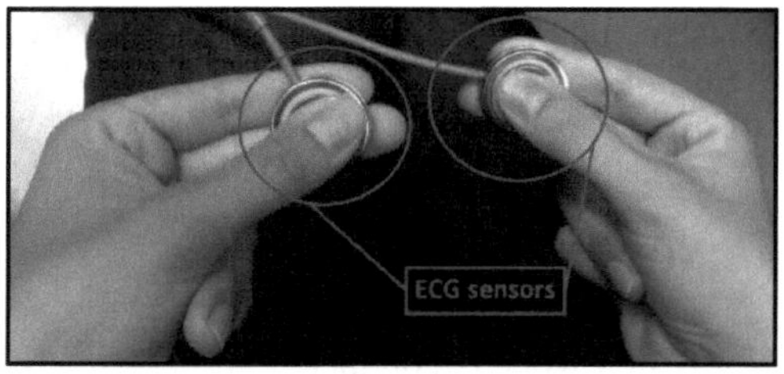

(a) EPIC sensors

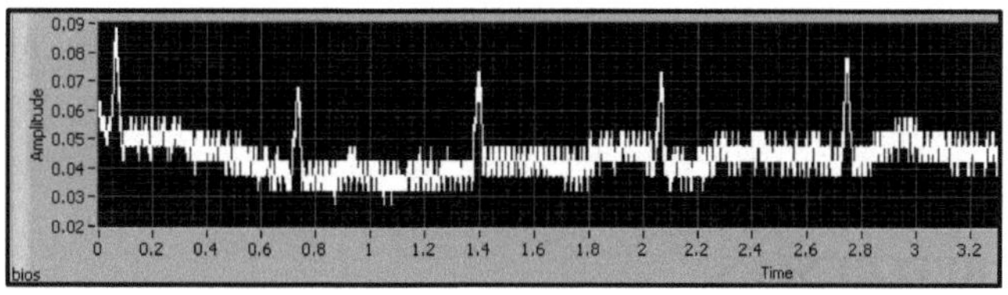

(b) Sample of ECG signal acquired

Figure 1. EPIC sensors for acquisition of ECG Lead II equivalent signals

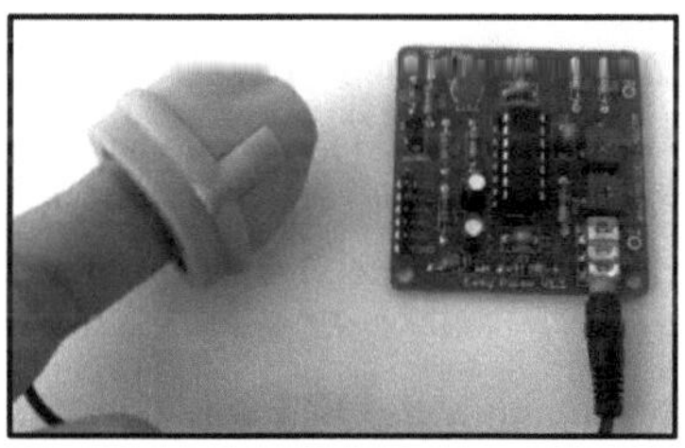

(a) Easy Pulse Sensor

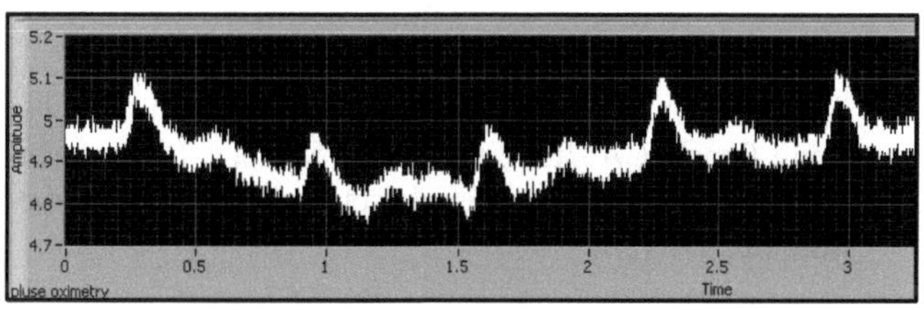

(b) Sample of PPG signal acquired

Figure 2. Easy Pulse sensors for acquisition of PPG equivalent signals

A. ECG and PPG pre-processing

Digital filters are used for the pre-processing of both ECG and PPG data as illustrated in Figure 3. The first filter is a 10th order infinite impulse response (IIR) filter with high pass filter with cut off frequency set at 0.5 Hz to eliminate low frequency noise. In the second filter, high frequency noises are removed by a low pass filter with cut off frequency set at 30 Hz. Finally, the last filter is a moving average filter used to reduce base-line wandering and smooth the ECG waveform for peak detection.

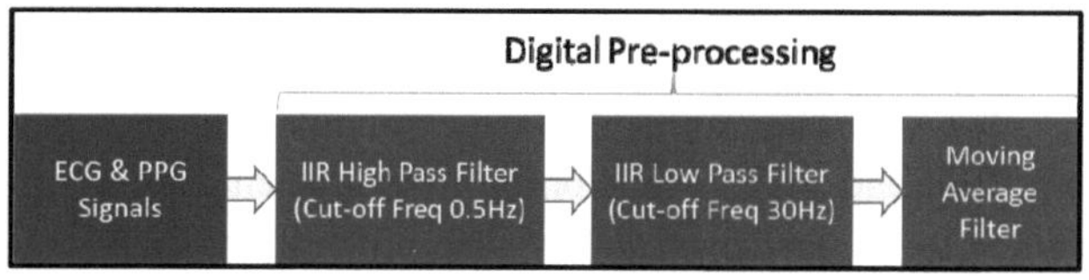

Figure 3. Digital Filtering of ECG and PPG signals

B. Adaptive threshold for peak detection

In both myRIO and BBB platforms, the threshold based peak detection function was used. The threshold value is originally set at half the supply voltage and is progressively reduced until the first peak is found. To achieve reliable peak detection, experiments were conducted to determine the optimal settings for the adaptive threshold voltage for peak detection. A cluster of five peak values were found to give repeatable and reliable peak detection. Equation 1 was derived based on these experiments and is used to determine the subsequent thresholds.

$$V_{threshold} = 0.7(V_{max}) - (2.776) * V_{std} \qquad \text{Equation 1}$$

where V_{max} = Maximum peak voltage from cluster of 5 peaks
V_{std} = Standard Deviation extracted from peak voltages

C. Cross comparison ECG vs PPG

Using the cardiovascular physiology that each of the cardiac cycle comprises an ECG followed by a PPG pulse, separated by a delay known as the pulse transit time (PTT), an algorithm is developed by cross comparing ECG and PPG peaks to enhance the accuracy of R-peak detection for ECG. For example, if an ECG peak is detected followed by the detection of a capillary pulse, the algorithm automatically identifies this as a valid ECG R-peak. If an ECG peak is not followed by a PPG peak, the algorithm automatically detects it as an extra ECG pulse which could possibly be a high T-wave detected by the adaptive threshold.

Through experimentations, it was found that a cluster of a five ECG and PPG peaks is able to give reliable and repeatable PTT. Equation 2 was derived from these experiments to define the lower and upper limits of PTT to differentiate ECG R-peaks from noise and artefacts.

$$Limits_{PTT} = Mean_{PTT} \pm (3.182/2) * Std_{PTT} \qquad \text{Equation 2}$$

where $Mean_{PTT}$ = Mean PTT from cluster of 5 peaks
Std_{PTT} = Standard Deviation PTT extracted from 5 peaks

III- RESULTS

The algorithm outlined above was implemented on both the BBB and myRIO platforms. Measurements were conducted on 6 volunteers where data were analysed in real-time and at the same time, stored. Post processing, using manual counting of peaks, was done to extract the accuracy of each implementations and the results are tabulated in Table 1 and Table 2.

Table 1. Measurement results using myRIO platform

Volunteer	Accuracy	Average accuracy
1	99%	98.6%
2	99%	
3	100%	
4	100%	
5	95%	
6	99%	

Table 2. Measurement results using BBB platform

Volunteer	Accuracy	Average accuracy
1	99%	96.5%
2	99%	
3	95%	
4	99%	
5	92%	
6	94%	

IV- CONCLUSIONS

Concurrent processing of both ECG and PPG is able to reduce the need to have high quality ECG and allows the use of simple signal processing algorithms to identify the locations of R peaks in ECG signals. This method is able to improve the accuracy in the detection of R-peaks in ECG with Signal to Noise Ratio (SNR) of 1-3dB, which is a challenge in existing methods due to the presence of noise; P and T-waves in ECG. The flexibility of our method was also demonstrated through concurrent implementation on a low cost platform (i.e. BeagleBone Black (BBB)) and FPGA platform (i.e. myRIO from National Instrument), achieving respective accuracy of 96% and 98%, using physiological signals acquired in real-time. The accuracy provided by our method is able to be applied on wearables and supports accurate real-time assessment of cardiovascular health.

V- ACKNOWLEDGEMENT

This research work is conducted at the Nanyang Polytechnic (NYP) of Singapore as part of the student exchange program between ESIEE, Paris and NYP, Singapore. The work is funded by the research grant from the Ministry of Education, Singapore.

REFERENCES

1. Xue Q, Hu YH, Tompkins WJ. Neural-network based adaptive matched filtering of QRS detection. IEEE Trans Biomed Eng. 1992;39:317–29.
2. Pan J, Tompkins WJ. A real-time QRS detection algorithm. IEEE Trans Biomed Eng. 1985;32:230–6.
3. Zhang F, Lian Y. QRS Detection based on multiscale mathematical morphology for wearable ECG devices in body area networks. IEEE Trans Biomed Circuits Syst. 2009;3:220–8.
4. Fraden J, Neumann MR. QRS wave detection. Med Biol Eng Comput. 1980;18:125–32.
5. Okada M. A digital filter for the QRS complex detection. IEEE Trans Biomed Eng. 1979;26:700–3.
6. Engelse WA, Zeelenberg C. IEEE Comput Cardiology. Long Beach, CA: IEEE Computer Society; 1979. A single scan algorithm for QRS detection and feature extraction; pp. 37–42.
7. Hamilton PS, Tompkins WJ. Quantitative investigation of QRS detection rules using the MIT/BIH arrhythmic database. IEEE Trans Biomed Eng. 1986;33:1157–65.

8. Keselbrener L, Keselbrener M, Akselrod S. Nonlinear high pass filter for R-wave detection in ECG signal. Med Eng Phys. 1997;19:481–4.
9. Suppappola S, Sun Y. Nonlinear transforms of ECG signals for digital QRS detection: A quantitative analysis. IEEE Trans Biomed Eng. 1994;41:397–400.
10. Dokur Z, Olmez T, Yazgan E, Ersoy OK. Detection of ECG waveforms by neural networks. Med Eng Phys. 1997;19:738–41.
11. Barro S, Fernandez-Delgado M, Vila-Sobrino JA, Regueiro CV, Sanchez E. Classifiying multichannel ECG patterns with an adaptive neural network. IEEE Eng Med Biol Mag. 1998;17:45–55.
12. Fernandez-Delgado M, Barro Ameneiro S. MART: A multichannel ART-based neural network. IEEE Trans Neural Netw. 1998;9:139–50.
13. Afonso VX, Tompkins WJ, Nguyen TQ, Luo S. ECG beat detection using filter banks. IEEE Trans Biomed Eng. 1999;46:192–202.
14. Coast DA, Stern RM, Cano GG, Briller SA. An approach to cardiac arrhythmia analysis using hidden Markov models. IEEE Trans Biomed Eng. 1990;37:826–36.
15. Poli R, Cagnoni S, Valli G. Genetic design of optimum linear and nonlinear QRS detectors. IEEE Trans Biomed Eng. 1995;42:1137–41.
16. Börjesson PO, Pahlm O, Sörnmo L, Nygårds ME. Adaptive QRS detection based on maximum a posteriori estimation. IEEE Trans Biomed Eng. 1982;29:341–51.
17. Sörnmo L, Pahlm O, Nygards M.-E. Adaptive QRS detection: A study of performance. IEEE Trans Biomed Eng. 1985;32:392–401.
18. Balda RA, Diller G, Deardorff E, Doue J, Hsieh P. The HP ECG analysis program. In: van Bemmel JH, Willems JL, editors. Trends in Computer processed Electrocardiograms. North Holland, Amsterdam, The Netherlands: 1977. pp. 197–205.
19. Ahlstrom ML, Tompkins WJ. Digital filter for real-time ECG signal processing using microprocessors. IEEE Trans Biomed Eng. 1985;32:708–13.
20. Friesen GM, Jannett TC, Jadallah MA, Yates SL, Quint SR, Nagle HT. A comparison of the noise sensitivity of nine QRS detection algorithms. IEEE Trans Biomed Eng. 1990;37:85–97.
21. Tompkins WJ. Biomedical digital signal processing. Upper Saddle River, NJ: Prentice-Hall; 1995.
22. Benitez DS, Gaydecki PA, Zaidi A, Fitzpatrick AP. A new QRS detection algorithm based on the Hilbert transform. Comput Card. 2000;1:379–82.
23. Chang WH, Lin KP, Tseng SY. ECG analysis based on Hilbert transform descriptor. IEEE Eng Med Biol Soc. 1988;1:36–7.
24. Chang AA, Chang WH. Hilbert transform and Fourier descriptors to ECG signal analysis. Proc IEEE Conf Eng Med Biol Soc. 1988;1:512–3.
25. Xu X, Liu Y. Adaptive threshold for QRS complex detection based on wavelet transform. 27th Annual Int. Conf Proc IEEE Eng Med Biol Soc. 2005;7:7281–4.

PSO based Density Classifier for Mammograms

Sneha Simon[1], R. Lavanya[2], Devi Vijayan[3]

[1]Department of Electronics and Communication Engineering, Amrita School of Engineering, Coimbatore, Amrita Vishwa Vidhyapeetham, Amrita University, India.
treesamampallil@gmail.com
[2]Department of Electronics and Communication Engineering, Amrita School of Engineering, Coimbatore, Amrita Vishwa Vidhyapeetham, Amrita University, India.
[3]Department of Electronics and Communication Engineering, Amrita School of Engineering, Coimbatore, Amrita Vishwa Vidhyapeetham, Amrita University, India.
devivijayan@gmail.com

Abstract—Breast cancer is the major cancer diagnosed in both, developed and developing countries. Early detection and treatment of breast cancer is necessary to moderate the associated fatality rates. Mammography is the widely accepted modality for screening breast cancer. Breast density is considered one of the major risk indicators for Breast cancer. Nevertheless, low contrast and subtle nature of abnormalities reduces the sensitivity of mammograms, especially in dense breast. In this paper we present an automatic method for breast density classification based on two level cascaded support vector machine (SVM) classifiers. Particle Swarm Optimization (PSO) has been employed for SVM parameter optimization that resulted in a low set up time for building the system. The proposed system was tested on mini-MIAS database, and an overall classification accuracy of 82% was achieved. Also the system could prompt the radiologists on high-risk cases, thereby gaining more attention from them for diagnosis of such cases.

Keywords - particle swarm optimization; mammogram; breast density.

1. INTRODUCTION

Breast cancer is the prime mover, for cancer related death in women. According to World Health Organization, this cancer has been a main health issue in both developed and developing countries. Although breast cancer is a fatal disease, primal detection can reduce the mortality rate. By now mammography is the most coherent tool for screening of breast cancer. Low contrast and subtle nature of abnormalities especially in dense breast render the radiologists finding to be subjective in nature. High breast density is lined with high risk of breast cancer [1]. Breast density classification also leads to false positive reduction so that unwanted biopsies can be eliminated [11].

This paper aims to classify breast tissues into different categories based on its density. In this paper a reliable automatic method for density classification of mammogram is investigated. Further it increases the scope for computed aided detection of abnormalities in order to assess the breast composition. The proposed work employs a combined SVM classifier for breast density classification. PSO has been employed for SVM parameter optimization that resulted in a low set up time for building the system.

Further the work is organized as follows. Part I impart the motivation and objective whereas, II covers the literature review related to this work. III describes the methodology and IV provides the results and discussion. Section V gives a conclusion to this work.

2. PREVIOUS WORK

Major works related to breast density classification is listed below. A novel density classification method[1] classifies the breast tissue into fatty and dense tissues. In this work Bayesian combination is used for classification, with textural and morphological features, were two data sets have been used. In this work both automatic and radiologist assessment, on Breast Imaging Reporting and Data System (BIRADS) showed high correlation. In[2], breast tissue classification is performed based on weighted voting tree scheme, which classifies the tissues based on textural

J. Goh et al. (eds.), *The 16th International Conference on Biomedical Engineering*,
IFMBE Proceedings 61,
DOI: 10.1007/978-981-10-4220-1_13

features. Here different classifiers combined together with linear discriminant analysis showed better performance. In [3], breast tissue is classified into dense, glandular and fatty using SVM classifier without optimizing the parameters. Statistical features of 43 normal mammograms were used, and obtained an accuracy of 95.44%. Work in [4] determines the mammographic risks, using a topographic map and breast density was estimated, and tissue classification was performed. In a recent work[5], breast tissue classification is performed using texture features. Here principal component analysis is carried out for feature selection, SVM and probabilistic neural network classifiers were used. In [6] an approach for tissue classification was proposed, where for different tissue type determines a distribution and based on this distribution each tissue type is classified. SVM and k nearest neighbor (kNN) were used as classifiers, and it showed SVM outperforms kNN.

Parameter selection is a crucial step in classification. Classifier with optimal parameters results in better classification. There are various approaches for parameter selection of SVM classifier. PSO, an optimization technique is one among the widely used method. In work[7], it shows that PSO been used for the parameter selection, as well as for feature selection. As density is highly lined with cancer, further fusion of different modality images can be utilized for high risk categories [10].

In the proposed work the parameters for SVM classifier is been optimized using PSO, which resulted in the improvement of generalization performance of the classifier.

3. METHODOLOGY

Breast tissue classification is done using various techniques. Mammogram images are taken, and it undergoes various image processing steps before classification. To improve the image quality, images undergoes various preprocessing techniques. From the pre-processed image features are extracted and then fed to the classifier, for the classification. The proposed work flow is depicted in Fig. 1

3.1 Preprocessing

Preprocessing techniques are performed prior to high level image processing to enhance the quality of mammogram image, aiding in higher classification accuracy. As mammograms are low contrast image, it undergoes preprocessing steps. Presence of noise deteriorates the quality of the image. Mammogram posses quantum noise. Wiener filter is applied for the noise removal, which calculates the variance and mean of the neighborhood pixels thereby smoothens the image.

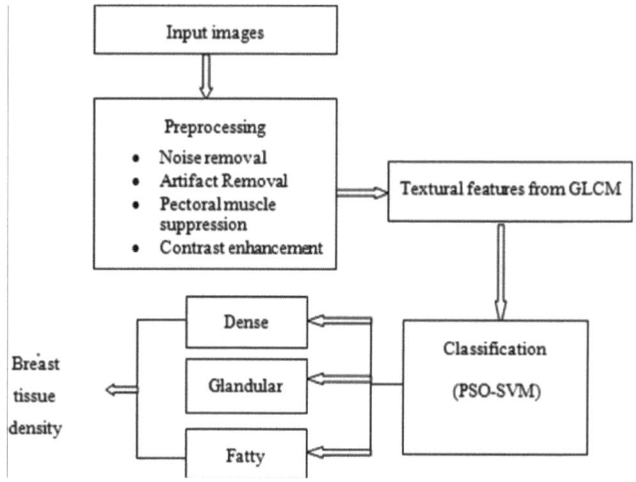

Fig.1. Proposed work flow

Table I. Extracted textural features

Sl.No.	Features
1	Energy
2	Contrast
3	Correlation
4	Variance
5	Sum average
6	Sum entropy
7	Sum variance
8	Entropy
9	Difference variance
10	Difference entropy
11	Correlation information 1
12	Correlation information 2
13	Homogeneity 1
14	Homogeneity 2
15	Cluster shade
16	Cluster prominence
17	Autocorrelation
18	Dissimilarity
19	Maximum probability

Being low contrast, it is always good for mammograms to undergo some contrast enhancement techniques. In this proposed work Contrast limited adaptive histogram equalization (CLAHE) is adopted in which the function operates on small regions of the images named tiles.

Features are extracted from the breast region, i.e., region of interest, other than that acts as an artifact. Artifacts are removed by thresholding and connected component labeling [3].

Pectoral muscles are present on mammograms, which are high density regions whose presence affects the density based tissue classification. Thus pectoral muscles are eliminated using Seeded region growing algorithm [8].

3.2 Feature Extraction

To obtain relevant information for the classification process, essential features are required. Total of 19 textural features [9] has been extracted for this purpose. Features are extracted from Gray-level co-occurrence (GLCM) matrix. Features are calculated considering pixel distance by 1 pixel position and oriented along different directions including 0 °, 45°, 90° and 135°. Extracted textural features are shown in Table I.

3.3 Classifier

In recent studies, SVM acts as an emerging classifier, which is widely used in different areas for classification purpose. In this work SVM Classifier is applied, for the breast tissue classification.SVM is a supervised machine learning algorithm based on the concept of maximizing the margin. Classifier is trained with a set of features, and based on this trained system, set of features are tested and classified.

In this work, breast density classification is performed by two-level cascaded SVM classifiers, in which the first SVM performs the classification of breast tissues into high and low dense tissues, whereas in the second SVM, low dense tissue is further classified into fatty and glandular. Both are non-linear SVMs that employ the radial basis function (RBF) kernel [5]. Choosing optimal parameters of the classifier has a great impact on classification accuracy. The RBF kernel parameters, C and gamma of SVM classifier are optimized by 10-fold cross validation.

k fold cross validation is a technique, were the total samples are partitioned randomly with equal k folds.k-1 folds are used for training and remaining for testing. This process repeats k times, thus each fold used as validation set at least once. In this work to obtain the optimal parameters for the classifier, k-fold cross validation is performed, with k value 10.

3.4 Optimization of parameter using PSO

PSO is an optimization technique which has similarity with the evolutionary algorithms. It is a population oriented technique. In an optimization problem, the particles represents the potential solution [7] and it search for an optimal solution by updating its velocity and position based on its fitness value. In this work PSO is used to optimize the C and gamma values of the SVM classifier. Cross validation accuracy acts as the fitness function in this problem. Objective behind the PSO optimization is to improve the computational cost of the system. Advantage of PSO is, it is simple to implement and faster comparing to other optimization techniques [7].

PSO is initialized with a group of random particles (solutions) and then searches for optima by updating generations. Algorithm defines each particle in the D-dimensional space as $X_i = (x_{i1}, x_{i2},....,x_{iD})$, where the subscript 'i' represents the particle number and the second subscript is the dimension, number of parameters defining the solution. The memory of the previous best position is represented as $P_i = (p_{i1},p_{i2},....,P_{iD})$, and a velocity for each dimension is independently established as $V_i = (v_{i1}, v_{i2},...,v_{iD})$. After each iteration, the velocity term is updated using equation 1, and the particle is moved with some randomness in the direction of its own best position, *pbest,* and the global best position, *gbest.* This is apparent in the velocity update equation, given by

$$V_{id}^{(t+1)} = \begin{array}{l} \omega\times V_{id}^{(t)} + U[0,1]\times\psi_1\times(p_{id}^{(t)} - x_{id}^{(t)}) + \\ U[0,1]\times\psi_2\times(p_{gd}^{(t)} - x_{id}^{(t)}) \end{array} \quad (1)$$

The position is updated using this velocity given in equation (2).

$$X_{id}^{(t+1)} = X_{id}^{(t)} + V_{id}^{(t+1)} \quad (2)$$

where U [0,1] samples a uniform random distribution, t is a relative time index, ψ_1 and ψ_2 are weights trading off the impact of the local best and global best solutions' on the particle's total velocity.

4. RESULTS AND DISCUSSION

As part of preprocessing, various techniques have been performed on mammogram images to enhance the quality of the image. Quantum noise is the significant noise that mammogram possess, and it was removed by applying Wiener filter. To quantify the performance of the filter it has been compared with median filter by estimating its PSNR value. Performances of two filters are given in Table 2. The PSNR value proves that Wiener filter outperforms median filter.

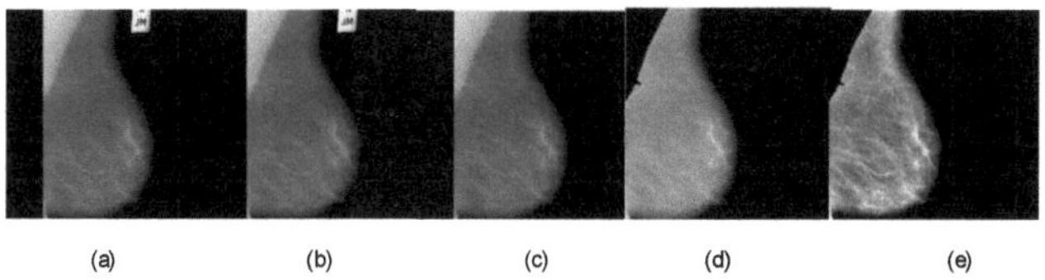

Fig. 2. Preprocessing. (a) input image (b) noise removal (c) label removal (d) pectoral muscle suppression (e) contrast enhancement

Table 2. Performance of filters

Image	PSNR for Wiener (dB)	PSNR for Median (dB)
mdb001	53.59	53.18
mdb002	51.60	49.01
mdb003	52.80	50.63
mdb004	52.64	52.35
mdb005	51.26	50.78
mdb006	51.04	50.90

The region of interest in the proposed work is breast region alone, hence labels present on mammogram acts as artifacts and it was removed by connected component labelling and thresholding. Pectoral muscles were removed by performing seeded region growing algorithm, and further the image undergone the contrast enhancement. Fig. 2 shows resulted images that underwent the mentioned preprocessing techniques.

19 textural features are extracted, and are fed to the SVM classifier. An obtained result shown in table 3 shows that, SVM classifier gives an overall accuracy of 82%. SVM-1 gives the classification results between low and high dense breast tissues. This classifier gives an accuracy of 86.82%, whereas SVM-2 that gives the classification results between the low dense breast tissues, which is further classified into fatty and glandular. It gives an accuracy of 81.81%.

During classification it is essential to have optimal parameters for a classifier. A nested two-level 10-fold cross validation is performed for parameter tuning and performance assessment. The RBF kernel parameters, C and gamma of SVM classifier are optimized by 10-fold cross validation. To avoid exhaustive search which is time consuming, especially for two classifiers, particle swarm optimization (PSO) is employed for the validation process. The proposed work emphasize on the improvement in the computational time

Optimization technique PSO was performed to optimize the SVM parameters. Optimization results are shown in the Table 4. It shows that computational time taken by 10 fold with PSO is considerably lesser, comparing with that of 10 fold cross validation without PSO. Overall accuracy of the classifier was maintained with PSO, which concludes PSO performed without compromising the classifier accuracy.

Table 3. Classification results.

Classifier	True positive	True negative	Total test images	Accuracy
SVM-1	16	96	129	86.82%
SVM-2	43	46	110	81.81%
Total accuracy of the classifier : 82%				

Table 4. Improved computational COST.

Classifier	Time taken without PSO	Time taken with PSO
SVM-1	106 hr	36 hr
SVM-2	47 hr	20 hr

5. CONCLUSION

In this work PSO-SVM framework is used for breast density classification. 10-fold cross validation was a time consuming process when implemented as an exhaustive search process of two cascaded SVMs. Thus PSO, an optimization technique was used to implement the validation process. Results showed that time taken for parameter selection of the classifier was considerably reduced. Parameter selection with k-fold cross validation and PSO were compared and analyzed, that PSO based classifier is better, which improved the computational cost by reducing the computational time. Images from Mini-MIAS database were considered in this work. Accuracy obtained by classifier without PSO was 82%, and PSO could achieve the same accuracy value with less computational time.

REFERENCES

[1] A. Oliver, J. Freixenet, R. Marti, J. Pont, E. Perez, E. R. E. Denton, and R. Zwiggelaar, "A Novel Breast Tissue Density Classification Methodology," *IEEE Trans. Inf. Technol. Biomed.,* vol. 12, no. 1, pp. 5565, Jan. 2008.

[2] N. Vallez, G. Bueno, O. Deniz, J. Dorado, J. A. Seoane, A. Pazos, and C. Pastor, "Breast density classification to reduce false positives in CADe systems," *Comput. Methods Programs Biomed.,* vol. 113, no. 2, pp. 569-584, Feb. 2014.

[3] T. S. Subashini, V. Ramalingam, and S. Palanivel, "Automated assessment of breast tissue density in digital mammograms," *Comput. Vis. Image Underst.,* vol. 114, no. 1, pp. 33-43, Jan. 2010.

[4] Z. Chen, A. Oliver, E. Denton, and R. Zwiggelaar, "Automated mammographic risk classification based on breast density estimation," in *Pattern Recognition and Image Analysis,* Springer, 2013, pp. 237-244.

[5] Kriti, J. Virmani, N. Dey, and V. Kumar, "PCA-PNN and PCA-SVM Based CAD Systems for Breast Density Classification," in *Applications of Intelligent Optimization in Biology and Medicine,* vol. 96, A.-E. Hassanien, C. Grosan, and M. Fahmy Tolba, Eds. Cham: Springer International Publishing, 2016, pp. 159-180.

[6] A. Bosch, X. Munoz, A. Oliver, and J. Marti, "Modeling and classifying breast tissue density in mammograms," in *Computer Vision and Pattern Recognition, 2006 IEEE Computer Society Conference on,* 2006, vol. 2, pp. 1552-1558.

[7] J. Li and B. Li, "Parameters Selection for Support Vector Machine Based on Particle Swarm Optimization," in *Intelligent Computing Theory,* Springer, 2014, pp. 41-47.

[8] D. N. Ponraj, M. E. Jenifer, P. Poongodi, and J. S. Manoharan, "A survey on the preprocessing techniques of mammogram for the detection of breast cancer," *J. Emerg. Trends Comput. Inf. Sci.,* vol. 2, no. 12, pp. 656-664, 2011.

[9] M. Haralick, K. Shanmugam, and I. H. Dinstein, "Textural features for image classification," *Syst. Man Cybern. IEEE Trans. On,* no. 6, pp. 610-621, 1973.

[10] R. Lavanya and N. Nagarajan, "Information fusion in CAD systems for breast cancer diagnosis using mammography and ultrasound imaging: a survey", *Journal of Artificial Intelligence,* vol. 7, no. 3, pp. 113-122, 2014.

[11] Lavanya, N.Nagarajan and M. Nirmala Devi, "False positive reduction in computer aided detection of mammographic masses using canonical correlation analysis", *Journal of Theoretical and Applied Information Technology,* vol. 59, no. 1, pp. 139-145, 2014.

Development And Evaluation of WEEG: A Wearable 8-Channel System for Brain Computer Interfaces

Tri Thong Vo

Biomedical Engineering Department, International University, Vietnam
vo.tri.thong@gmail.com

Abstract. Research-grade Electroencephalogram (EEG) instruments offer the brain's electrical activity recording with high accuracy and a wide range of functionalities. Due to rigorous component requirements, these devices are usually expensive and elaborate. Recent advancements in electronic enable smaller designs to achieve comparable measurement results with similar functions such as impedance measurement. In this paper, we present the design of WEEG –A Wearable 8-channel System EEG recording device with integrated hardware and software interface. The circuit board which is smaller than a credit card size contains a power management system, an ADS1299 Analog to Digital Converter (ADC) to capture brainwaves and a microcontroller to handle a part of digital processing tasks. It communicates with a host computer via Bluetooth/Serial. Extensive evaluations were deployed to evaluate the performance of WEEG. We compared the signal quality of WEEG to that of the Biosemi Active Two -a conventional high-quality EEG recording system. Results came out comparable, indicating a high accurate recording of the proposed system. Also, Steady State Visually Evoked Potential (SSVEP) based game was devised. The high performance of users in the game suggests usage of our device in a wider range of Brain Computer Interface applications.

Keywords: Wireless EEG recording, Low-cost device, Brain-Computer Interface, System validation.

1 Introduction

The design of a low-cost and wireless EEG recording system so called was described by Vo et al in a presentation at BME6[1]. According to Fig. 1, the platform consists of a power management module, the microcontroller-wireless module and the analog to digital converter module. The ADS1299 is the central component of the sysstem because it determines overall specifications[2].We also have a peripheral controller STM32F4 that takes recording data from the Analog to Digital converter via SPI interface. The data package consists of 3 status byte information and 24 EEG data bytes. The microcontroller converts these two complement values to decimal and repackages them. The counter is added to these packages as a rudimentary way to monitor data integrity. A checksum or other techniques may be implemented to improve this process.

After processing the data, the STM32F407 chip sends data in packages to the computer via serial/Bluetooth for real-time display and data recording. Users may access these data via Bluetooth. This is system is performing at 380400 baud rate with a reliable output. During our investigation to use Bluetooth Low Energy (BLE) for the design, we find out that this protocol data transfer rate is slow, and the package size is limited to 20 bytes or less. It opposed to high data transfer rate of EEG recording devices whereas multiple channel information is recorded simultaneously. In the next stage, data will be processed within the STM32F407 chip on the device. It will be possible to use BLE to send status data to a smartphone or computer.

This design has been upgraded and extensively validated in order to improve the performance. In this paper, we present these design upgrades and system evaluations.

2 Update on overall specification

The most notable change was the system miniaturization. In the last modification, the circuitry was optimized, resulting in the total size of 6.5cm x 4 cm of the main board. **Error! Reference source not found.** shows the difference in size between the last version and the upgraded one. The minimized size of the current prototype provides versatile portability that traditional EEG recording systelms lack of. Another improvement was on the transfer rate. The firmware and the corresponding data-receiving software were enhanced to boost the speed of communication up. Specifically, the device reaches a remarkable wireless transfer rate at 500 Sample Per Second (SPS) for simultaneous eight channels. This data rate exceeds the speed of other consumer grade EEG devices. Other applications usually transfer at 250 SPS or lower.

The hightransferrate providesa powerful meantoinvestigate the brain'sactivitiesin new dimensions. For example, onset epilepsy seizure detection requires high EEG freq uency detection[4]

J. Goh et al. (eds.), *The 16th International Conference on Biomedical Engineering*,
IFMBE Proceedings 61,
DOI: 10.1007/978-981-10-4220-1_14

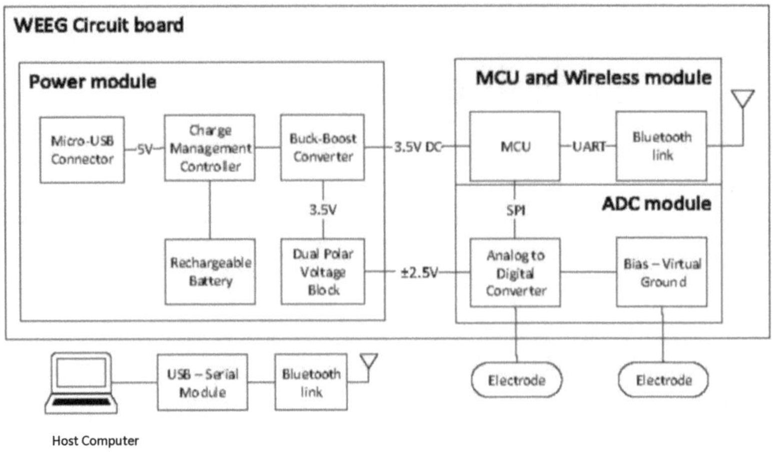

Fig. 1 WEEG system block Diagram

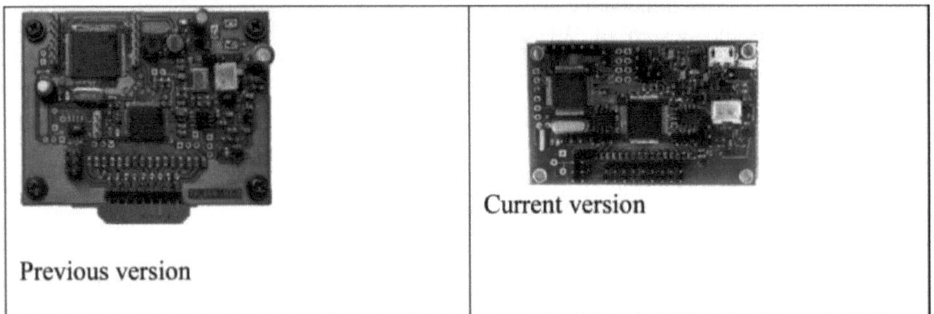

Fig. 2. Main board's size comparison between last version and upgraded one

Table 1 General specification comparision[3]

System	Manufacturer	Sampling rate (Hz)	Bandwidth (Hz)	Wireless transmission	Bitrate
WEEG	BME-IU	500	0.3-200	Bluetooth	24 bits
B-Alert X10	ABM	256	0.1–100	Bluetooth	16 bits
EPOC	Emotiv	128	0.2–43	Proprietary wireless	16 bits
HMS	QUASAR	240	0.02–120	Proprietary wireless	16 bits

This prominent feature can also enables more advanced Brain Computer Interface (BCI) applications. The high rate of transferring data allows designing highly responsive BCI systems, especially those that favors temporal features such as Event-related Potential (ERP) ones. Nonetheless, in a more general perspective, high transfer rate benefit all BCI applications since more data transferred in a shorter period of time means less delay during operating process, which yields better user using experience. This characteristic of is even more crucial in applications involving learning and training using EEG signals.

3 System Validation

3.1 Comparisons with Biosemi's ActiveTwo

The Biosemi's ActiveTwo is a well-known "gold standard" system among researchers.

Compare Power Signal Densities (PSD) at alpha band: The PSD of the brain are measured by both the WEEG and Biosemi's ActiveTwo during eye closed for comparison. Electrodes of WEEG are placed the same as Biosemi's ActiveTwoset up. As shown in Fig. 3, the recorded channel is at O2, the Common Mode Sense (CMS) and Driven Right Leg (DRL) of both system are at the same location near Pz.

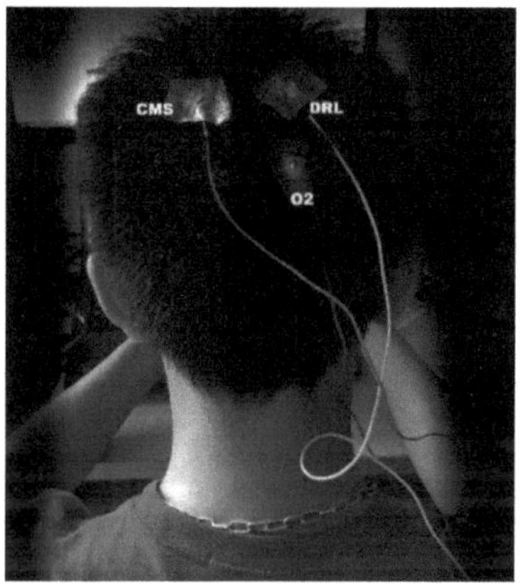

Fig. 3. Electrode placement for comparison with Biosemi's ActiveTwo

The experimental protocol is presented as follows. Two healthy subjects (all male, age 27 and 21) participated in the study. None had a history of diabetes, epilepsy, or any other chronic disease and were not taking regular medication. Subjects sit and relax on a chair with his eyes open for 45s. There are none stimuli in front of the subject's eyes. Second, the subject closes his eyes in 45s. This protocol is illustrated in Fig. 4. One run consists four trials. Gelled Electrodes are placed on the subject's scalp according to the Biosemi head cap. One channel is recorded at O2 location. One Ground (bias) electrode is placed on the DRL location of the cap. The differential pair is placed on the O2 and the CMS location. The sampling rate is 250 Hz for WEEG and 256Hz for ActiveTwo. EEG data was high-pass filtered at 0.3Hz to avoid DC drift and a low-pass filter at 43Hz for both systems.

Use Welch's method to estimate power spectral density. 20 second of data during eye closed are analyzed. The data are divided into 8 segments with 50% overlap. Each section is windowed with a Hamming window. Four epochs are averaged out to get the final mean. The WEEG signal is normalized for a straight forward comparison. The result is demonstrated in Fig. 5.

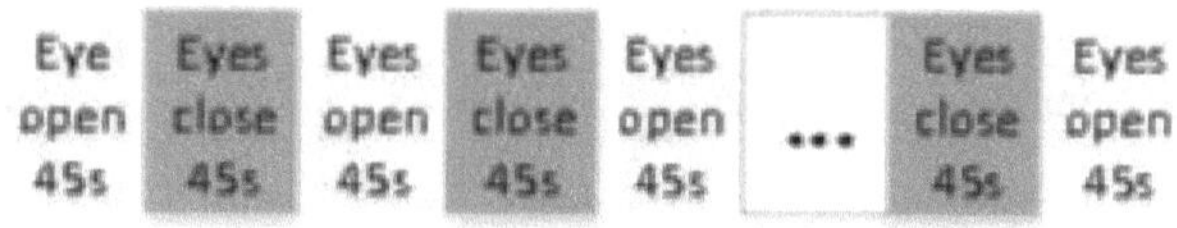

Fig. 4. Alpha waves comparison protocol

The WEEG's signal in blue is similar to the Biosemi's signal in red. Power in the alpha band from 8 Hz to 13Hz is significantly higher than other frequencies. Both graphs roll off gradually from 15Hz toward 22Hz. Higher frequencies' powers are notably trending down steeply. The normalized cross-correlation between two signal is very high at 0.9

Compare SSVEP PSD: This experiment is set up based on the SSVEP phenomena, a common technique in BCI. When stimulus (e.g flickering screen) is presented to the subject at a certain frequency, his/her brains wave will boost up at this frequency significantly.

We compared the PSD of the brain signals measured by the WEEG and by the Biosemi's Activetwo during stimulation at a fixed frequency 6.6Hz.The goal of this analysis is to evaluate the sensitivity of the system in measuring particular brain waves. Compared to the alpha wave, the SSVEP signals are much harder to detect, a qualified EEG recording device should able tocapture these types of EEG's pattern.

The subjects, who were introduced in the above section, sit on a chair and stares at the LCD screen. The flickering screen will alternate between stimuli (showing flickering at 6.6Hz) for 10s and non-stimuli (showing black screen) for 10s. Gelled electrodes are placedonthe subject'sscalp. One channel is recorded at O2 location. The DRL of the WEEG is put at the left mastoid, where as the reference electrode is placed at Oz. The sampling rate is 250 Hz for WEEG and 256Hz for ActiveTwo. EEG data was high-pass filtered at 0.3Hz to avoid DC drift and a low-pass filter at 43Hz for both systems.

Data recorded by the Biosemi's ActiveTwo is normalized from 256 SPS to 250SPS for comparison. Ten seconds of measurement data from both systems were analyzed. The magnitude squared coherence spectrum were calculated by Matlab function mscohere.

According to the definition of the term, the magnitude-squared coherence estimate is a function of frequency with ranges between 0 and 1 that indicates how well x corresponds to y at each frequency. The magnitude-squared coherence is a function of the power spectral densities, Pxx(f) and Pyy(f), of x and y, and the cross power spectral density, Pxy(f), of x and y.

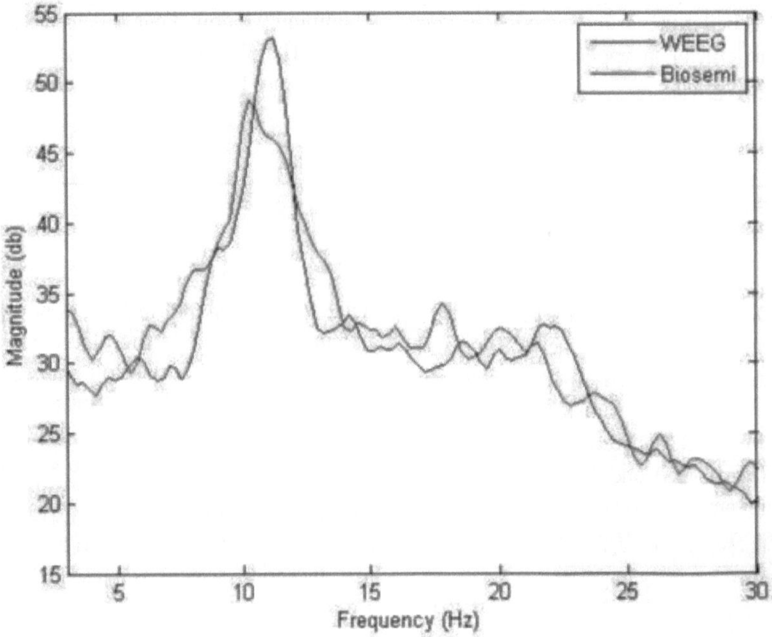

Fig. 5. A comparison of PSDs for WEEGandBiosemi recordings for the "eyes closed."

$$C_{xy}(f)\frac{|P_{xy}(f)|^2}{|P_{xx}(f)|\cdot|P_{yy}(f)|}$$

At target frequency and its harmonics, there is strong coherence (> 0.7) between data sets from both systems (Fig. 6). It proves that the WEEG can produce meaningful data for the certain experiment.

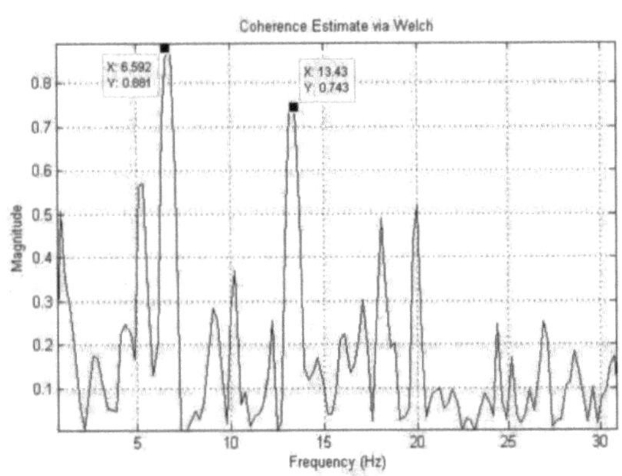

Fig. 6. Coherence Estimation via Welch's method

3.2 SSVEP BCI Application – A Cursor Game

We extended the SSVEP experiment to a full Brain-Computer Interface application which consisted of 4 different stimuli. A 26-year-old subject sits on a chair and stares at the stimulus LCD screen. The flickering screen will alternate between stimuli (showing flickering at 6.6Hz, 7.5Hz, 8.75Hz, and 10Hz) for 10s and non-stimuli for 10s as shown in Fig. 7. We record four trial in a run. The subject looks at one of these frequencies, 6.6Hz, 7.5Hz, 8.75Hz, and 10Hz, respectively for every trial. Electrodes placements and filters setting are the same as the previous experiment.

Using SSVEP, we can design a system that allows users select different targets, for example, navigating an electrical wheelchair. In this experiment, the user played a cursor game in which he had to control the Super Mario avatar and navigated it into the location of the Princess icon (Fig. 8). The subject was able to move the avatar at the rate of 3.3 second/movement.

Fig. 9 displays the spectrogram of the recording. The window for the spectrogram calculation is 2000; window overlap is 80%; sampling rate is 250 sample/second. Red marks in the spectrogram represent high power area where as blue marks denotes low power. Long red stripe at target frequencies (6.6Hz, 7.5Hz, 8.75Hz, and 10Hz) indicate Evoked Potential of the brain.

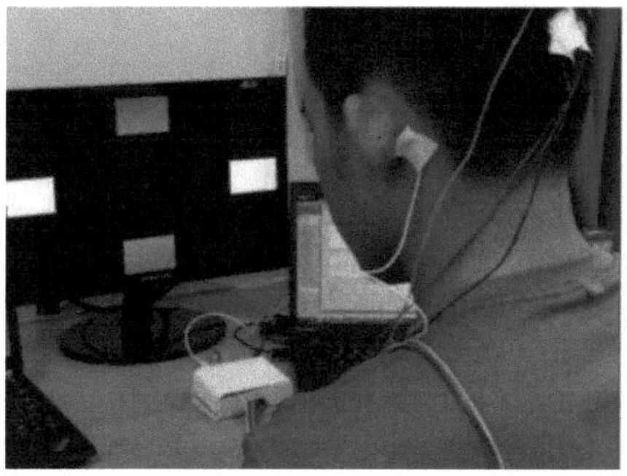

Fig. 7. Illustration of a SSVEP test session

Fig. 8. The Cursor Game Interface

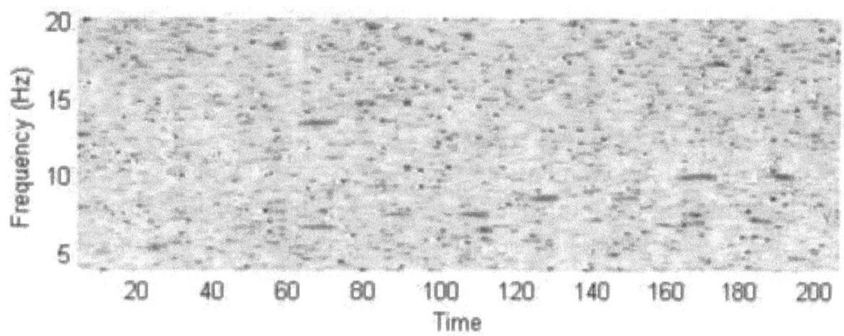

Fig. 9. Spectrogram of an SSVEP session

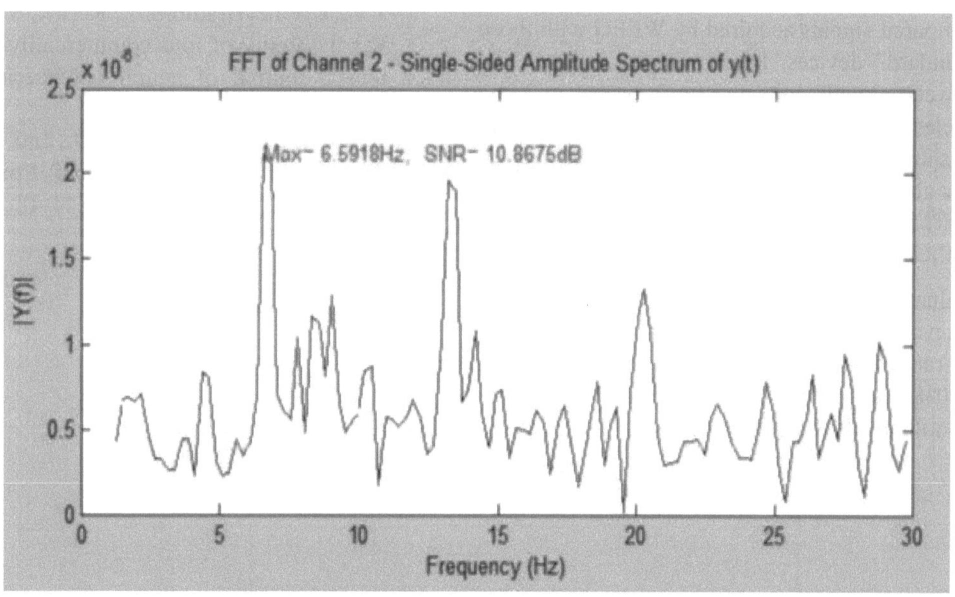

Fig. 10. Amplitude Spectrum during stimulation

Fig. 10 illustrates the FFT -Amplitude Spectrum of recorded brainwaves when the stimulation is at 6.6 Hz. The window length is 512 data point; zero padding to 1024 FFT point for better frequency resolution. The figure shows three majors peaks at the base frequency 6.6Hz, harmonics at 13.2Hz and 20.26Hz. During a good recording run, the mean Signal to Noise Ratio (SNR) at stimulation frequencies is 13 dB. A typical mean SNR is at 10dB. These conclusions are withdrawal from 5 runs at the different target frequency. The SNR is calculated based on the equation below:

$$SNR = 20\log\frac{Signal}{Noise}(db)$$

, where:

- Signal is the amplitude at the stimulus frequency
- Noise: is the average amplitude of frequency within the interest range from 5Hz to 25Hz

4 Conclusion

In this work, updates on the WEEG – wireless EEG recording system is presented. Two major improvements are miniturization of the main board, resuting in the total size of a credit card. The other is the increase in data-transferring rate, up to 500 samples per second for 8 channels recorded at the same time.

Another contribution of this study is to provide an extensive evaluation on WEEG, establishing an original reference of the system's quality and capabilities. In the signal quality testing, we compared signals acquired by WEEG with those of a "gold standard" devices, Biosemi's ActiveTwo. The coherence between signals collected by 2 system at some target frequencies is higher than 0.6, which is comparable. This analysis shows that the proposed system can acquire signal that is as good as the traditional one. In some applications, the WEEG is also a better choice because of its portability and simplicity.

In another evaluations, a SSVEP-based BCI game was devised to demonstrate WEEG system's capability of performing wider ranges of applications. With the total accuracy was up to 90%, the system was proven to be highly functional and versatile in a broad band of usage.

5 Discussion

This novel platform paves the way to develop endless medical devices based on EEG signal. Between conventional hundred thousand dollar system and not having anything at all are a huge gap to fill in. Unequipped hospital in remote areas will be beneficial from these new low-cost devices. The Community Health Medical Center, in Ho Chi Minh City, Vietnam, expresses their interest to use this platform as a sleep monitoring device at patients' home. Abundant requests to diagnose sleep disorders contrast with limited equipment in the center. This device can play a vital role to alleviate similar pressing issues.

References

1. Vo Tri Thong, N.P.N.a.V.V.T.: 'WEEGEE: Wireless 8-Channel EEG Recording Device', The 6th International Conference on the Development of Biomedical Engineering in Vietnam, 2016
2. TexasInstruments: '"Low-Noise, 8-Channel, 24-Bit Analog Front-End for Biopotential Measurements" ADS1299 Datasheet', 2012
3. David Hairston, W., Whitaker, K.W., Ries, A.J., Vettel, J.M., CortneyBradford, J., Kerick, S.E., and McDowell, K.: 'Usability of four commercially-oriented EEG systems', Journal of neural engineering, 2014, 11, (4), pp. 046018
4. Jouny, C.C., Franaszczuk, P.J., and Bergey, G.K.: 'Improving early seizure detection', Epilepsy & behavior : E&B, 2011, 22 Suppl 1, pp. S44-48

Performance Analysis of GMM Classifier for Classification of Normal and Abnormal Segments in PPG Signals

Sunil Kumar Prabhakar, Harikumar Rajaguru

Department of ECE, Bannari Amman Institute of Technology, India

Abstract. Photoplethysmography (PPG) is widely used to estimate the blood flow of skin by utilizing the infrared technique. Parameters such as blood pressure, oxygen saturation levels, blood saturation levels and cardiac output levels can be measured easily. As PPG is non-invasive in nature and it has a low production and maintenance cost, it is widely used in clinical practices. The performance analysis of Gaussian Mixture Model (GMM) as a post classifier is utilized in this paper for the classification of normal and abnormal segments in PPG Signals. The main objective of the paper is to identify normal and abnormal PPG Segments of the PPG waveform observed in the long time monitoring of the Physionet Data Base available online for a particular patient. The PPG Signals are sampled at 100 Hz. The PPG data sample length obtained is 1, 44, 000 and it is segmented into equal intervals comprising of 200 samples totally. Therefore the entire data consists of 720 segments. Totally ten different features such as mean, variance, standard deviation, skewness, kurtosis, energy, approximate entropy, peak maximum, maximum slope, and Singular Value Decomposition (SVD) are extracted and normalized. Based on the SVD values, each segment is labeled as normal or abnormal segment. The normalized features are given as inputs to the GMM classifier to classify the normal and abnormal segments in the PPG Signals. The performance metrics analyzed in this work are specificity, sensitivity, accuracy, precision and False Discovery Rate (FDR). Results show that an accuracy of 98.97% is obtained, precision of 100%, nil FDR, specificity of 100% and sensitivity of 97.95% is obtained.

Keywords: PPG, GMM, SVD, Normal, Abnormal

1 Introduction

PPG is a non-invasive technique which has a very low cost and it is an easy maintenance tool for the purpose of monitoring the changes in the blood volume incessantly in the tissues with respect to function of time [1]. It is one of the standard non-invasive techniques for the measurement analysis of oxygen saturation level and has been widely acknowledged by both European committee for standardization and International Standard Organization (ISO). In various clinical fields like critical care unit, surgical recovery units and in anesthesia, PPG is used efficiently. Due to its portable and non-intrusive nature, the techniques involved with PPG have been a subject of extensive and in-depth research in the recent decades. With drastic changes and improvements emerging in the field of clinical instrumentation, Digital Signal Processing, opto electronics and so on PPG has achieved a versatile spectrum of different applications initiating from vascular assessment, physiological monitoring in a clinical manner and evaluation of autonomic functions [2]. During the measurement process, the PPG signals can be affected easily which paves the way for inaccurate analysis of PPG waveform. Some of the major errors raised in PPG are due to the ambient light source at the photo detector or the peripheral tissues which can have a poor blood perfusion and the effect of motion artifacts. In environments such as home care settings, these potential errors are very frequently acting as obstacles for the reliable usage of PPG derived parameters in both continuous and real time monitoring applications. So it is necessary to provide a trust metric of the signal quality so that with the in-depth analysis of algorithms, a high level of trust can be obtained in the parameters which are desired.

In the medical field analysis, finger PPG is a commonly used technique because of its various applications and utilities. It is based on the strong determination of the optical properties of a particular area of skin. Inside the skin, infrared light which is non-visible is emitted. Depending on the volume of the blood in the skin, more light can be absorbed or less light can be absorbed. The light which is back scattered corresponds only to the different variations of the volume of blood. The determination of the changes of blood volume can be easily done by measuring the light which is reflected, thereby making complete use of the optical properties of blood and the skin tissue. The most wonderful applications of the PPG waveform analysis is that it provides a rapid biophysical measure of various diseases and aging process [3]. An approximate estimation of coetaneous blood flow is provided by the PPG. The dynamic

J. Goh et al. (eds.), *The 16th International Conference on Biomedical Engineering*,
IFMBE Proceedings 61,
DOI: 10.1007/978-981-10-4220-1_15

attenuation of infrared light is measured by PPG depending on the volume of blood present in the tissue. If the PPG's contour is analyzed in the context of crest time, it can reveal the age dependence and the most significant clinically varying health diseases such as hypertension, arteriosclerosis and so on. In PPG, the arterial pulsations form an important aspect. The oscillatory component of the PPG describes a pulsatile wave, whose respective contour provides useful description and content about the vascular health. The waveform of PPG has a pulsatile (AC) physiological waveform and with each heart beat, it is attributed to cardiac synchronous change in the volume of blood. The PPG measures are done in the index fingers as it is quite easier for recording. The pulse shape has been characterized by the quantitative measures and the derivatives of it are highly useful for the analysis of PPG signal. For the assessment of various health diseases, the peripheral pulse is commonly used.

Some of the related works in the PPG signal analysis is as follows. Allen reviewed in a detailed manner about the PPG and its applications in clinical physiological measurement. The signal quality measures for pulse oximetry through waveform morphology analysis were performed by Sukor et al [4]. A computational system to optimize noise rejection in PPG signals during motion or poor perfusion states was developed by Foo and Wilson [5]. Gray Beal and Petterson showed that adaptive filtering and alternative calculations revolutionizes pulse oximetry sensitivity and specificity during motion and low perfusion [6]. Yan explained the reduction of motion artifact in pulse oximetry by smoothed pseudo Wigner-Ville distributions [7]. The motion artifact reduction in PPG using ICA was developed by Kim and Yoo [8]. The aortic and arterial stiffness determination by PPG techniques was done by Huotari et.al [9]. The Photoplethysmographic characterization of the vascular wall by a new parameter called minimum rise-time was developed by Gavish [10]. The utility of the photoplethysmogram in circulatory monitoring was developed by Reisner et al [11]. The Artificial Neural Network (ANN) approach to PPG signal classification was done by Mohamed et.al [12]. The analysis of the effect of ageing on rising edge characteristics of the PPG using a modified Windkessel model was done by Edmond et al [13]. The Photoplethysmographic analysis of artery properties in patients with cardiovascular diseases was done by Rubins et al [14]. The PPG peaks analysis in patients presenting with erectile dysfunction was performed by Qawqzeh et al [15]. Validation of a device to measure arterial pulse wave velocity of a Photoplethysmographic method was done by Loukogeorgakis et al [16]. Photoplethysmographic augmentation index as a non-invasive indicator for vascular assessments was done by Gonzalez et al [17]. In this paper, the normal and abnormal PPG segments are classified with the help of a Bayesian based GMM model for a single patient suffering from respiratory disorder. The block diagram of the work is shown in Figure 1. The section methodology of the paper is as follows. In section 2, the materials and methods are discussed followed by the feature extraction and labeling of SVD in section 3. Section 4 explains in detail about the usage of GMM as a validator for classification purposes followed by results and discussion in section 5. The paper is concluded in section 6 followed by the references.

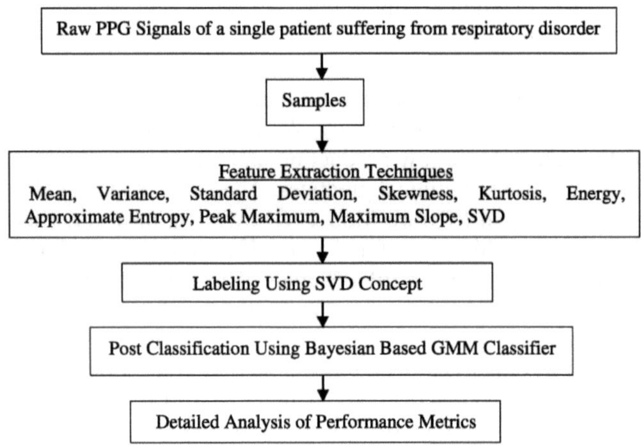

Figure 1 Illustration of the Work

2 Materials and Methods

The data has been obtained from the Physionet Database (MIMIC II waveform database) available online [18]. Thousands of recordings of multiple physiologic signal waveforms and the time series representation of important signs called as numerics are collected from bedside patient monitor screens for neonatal adult Intensive Care Unit (ICU) in this database. It is closely related to the MIMIC II Clinical Database, which has a lot of useful clinical information for many patients represented in the database of waveforms [19]. Depending on the numerous choices done by the ICU Staff, the numerics and recorded waveforms vary. The waveforms mostly include one or few ECG signals, fingertip PPG signals, arterial blood pressure waveforms, respiratory waveforms and various other waveforms. In the numerics, various measures like heart and respiratory rates, SpO2, systolic blood pressure, diastolic blood pressure and mean are together available. The length of the recordings can vary greatly, some recordings take a few days duration and some take several weeks, some are shorter and some are longer. The reading of a single patient who is suffering from respiratory disorders is efficiently used in this study analysis.

Generally, the PPG signal data acquisition procedure is as follows. Initially, in the transmitter mode PPG system, as a light source ultra-red LED of about 660 nm is used and as a photo detector, filter, amplifier and ADC, light dependent resistor (LDR) is used. The finger probe should comprise of LED and LDR that is mounted in an enclosure and can easily fit over the tip of the index finger of the patient. The finger probe should have very good contact to the surface of the skin and their respective optical axes should be parallel to each other. The LED and LDR are highly sensitive and has a rapid response and can be easily matched in their compactness and wavelength. If the power is on, the LED emits the light and it transmits through the finger of the patient. The photo detector is utilized to detect the transmitted light which is unabsorbed. With the periodic contraction of the heart, at the extremities, the blood is forced and the total amount of blood in the finger increases. The optical density is altered so that the transmission of light through the finger reduces and the voltage present in the output circuit of the photo detector changes significantly because of the changes in resistances of LDR. Using an amplifier, the low magnitude voltage is amplified with a suitable gain. To reduce the size of the dominant DC component, a high pass filter is also used. It also allows the boosting of pulsatile AC component to reach a nominal peak to peak level. The PPG signals can be later further digitized, monitored and stored safely in a system with the help of dedicated Data acquisition system which should have a built in ADC.

3 Feature Extractions and Annotation Using SVD

In the context of pattern recognition, signal and image processing, machine learning etc. feature extraction aims to get a set of non-redundant and informative data from an initial set of any considered data [20]. Building or developing derived values is the main goal of feature extraction. It always leads to a better interpretation by humans. When an input data is too large for processing an algorithm, and if it is suspected to contain a lot of redundant factors, and then feature extraction techniques can be employed to transform the input data into a particular set of reduced values named as features. If the complex data is considered for performance analysis, then a major chunk of problems starts from the total number of variables involved in it. If a larger number of variables are to be analyzed, it definitely requires a lot of memory, computational time and power. Also over fitting problems occurs easily in the classification algorithm. So it is necessary to extract the features of the PPG signals.

3.1 Feature Extraction Parameters:

Some of the features extracted here are mean, variance, standard deviation, skewness, kurtosis, energy, approximate entropy, peak maximum, maximum slope and SVD.

a) Arithmetic Mean:

To compute the arithmetic mean of a particular set of data, all the values of data represented as x should be added first and then it should be divided by the total number of values (n). The values that are to be summed is indicated by the symbol Z and so the formula for the mean (x) is represented as

$$\bar{x} = \frac{\Sigma x}{n}$$

b) Variance:

In the theory of probability and statistics, variance is the expectation value of the squared deviation of a random variable from its mean. It informs how far a particular set of randomly generated numbers are spread away from the mean.

The mathematical definition is represented as

$$Var(X) = E(X^2) - E(X)^2$$

Some of the properties of Variance are as follows

$$Var(c) = 0,\ Var(cX) = c^2 Var(X),\ Var(X + c) = Var(X)$$

c) Standard Deviation:

As the variance is measured in terms of x^2, the standard deviation is represented as $\sigma = \sqrt{Variance}$. Unlike variance, the standard deviation is measured in the same units as applicable to the original data.

d) Skewness:

A fundamental and major task in many statistical analysis techniques is to characterize the two input parameters clearly namely the variability and location of the data set. If we need to characterize the data further, then skewness can be analyzed. The variations of the data set are known with the help of measure of dispersion. The direction of the variations of the data set is given by skewness. It is a measure of symmetry or to be more precise, the lack of symmetry. A dataset or a distribution is said to be symmetric if it looks identical to both the sides, left and right if the centre point. For an univariate data $Y_1, Y_2, \ldots, Y_N$, the skewness is represented as

$$\text{Skewness}=\frac{\Sigma_{i=1}^{N}(Y_i-\bar{Y})^3/N}{S^3},$$

Where Y is denoted as mean, S represents the standard deviation and N represents the total number of data points.

e) Kurtosis:
It is a measure of the "tailedness" factor of the given probability distribution for a real time variable which is random in nature. The shape of a probability distribution is given by Kurtosis. For an univariate data Y, Y,--, Yr, the Kurtosis is given as

$$\text{Kurtosis}=\frac{\Sigma_{i=1}^{N}(Y_i-\bar{Y})^4/N}{S^4}$$

Where Y is denoted as mean, S represents the standard deviation and N represents the total number of data points.

f) Peak Maximum:
A local peak in a data sample that is largest than its neighbouring samples represent the peak maximum value.

g) Slope:
The slope of a particular line is nothing but a number that explains both the steepness and directions of the line. It is often denoted by the alphabetical letter$m$. The letter m is usually denoted for slope, because m represents multiple in the equation of straight line as

$$y = mx + c$$

The direction of a line can be decreasing, increasing, vertical or horizontal. The slope is called positive, if the line is increasing when it goes up from Left to Right (m > 0). The slope is called negative, if the line is decreasing going down from Right to Left (m < 0) . The slope is undefined if the line is vertical in nature and if the slope is zero, the line is horizontal in nature and this indicates a constant function.

h) Energy:
The energy of a discrete-time signal $x(n)$ is given as

$$E_s = \langle x(n), x(n) \rangle = \sum_{n=-\infty}^{\infty} |x(n)|^2$$

i) Approximate Entropy:
It is used for the exact quantification to assess the amount of regularity and fluctuation unpredictability over a given time-series data. The value of Approximate Entropy is quite low for regular time series data and the value of Approximate Entropy is high for irregular time series data which is quite complex.

j) SVD:
As a feature extraction technique, it is used in applications like finding pseudo inverse, total least squares minimization, rotation and scaling purposes, finding orthonormal bases and solving homogeneous linear equations. Supposing Q is considered as a $q\ x\ r$ matrix, where the entries of it has come from a field L , then there exist a factorization technique called SVD of Q, of the form

$$Q = A\Sigma B^*$$

where A denotes $q\ x\ q$ unitary matrix. Z represents the diagonal of $q\ x\ r$ matrix which has real numbers which are non-negative on the diagonal and $B = r\ x\ r$ unitary matrix over L . B^* is the conjugate transpose of the r x r unitary matrix B .

3.2 Annotation of SVD:

The labeling annotation of SVD is done for the easy classification purposes [21]. The threshold values are fixed as 60 to identify the normal and abnormal segments due to the respiratory problem. If the Eigen values of the SVD are more than 60, then it is denoted as abnormal value because of the respiratory problem and if the value of SVD is less than 60, it shows that the value is perfectly normal. The labeling of SVD as shown in Table 1 is done with the help of a physician and set as a gold standard.

Table 1 Annotation of PPG Signals Using SVD

Parameters	First Segment	Second Segment	Third Segment	Fourth Segment	Fifth Segment	Sixth Segment
Mean	1.93	0.042	0.3412	5.604	4.692	1.4148
Variance	12.19	11.148	13.096	21.986	25.784	10.2588
Std.Deviation	3.49	3.3388	3.618	4.689	5.077	3.2029
Skewness	0.63	0.6963	0.3912	0.7217	1.019	0.6219
Kurtosis	0.875	0.7909	0.905	0.7771	0.4902	0.8663
Energy	3176.11	2218.85	2629.58	10656.36	9534.088	2441.83
App. Entropy	1.59	1.4597	1.639	0.704	0.768	1.4706
Peak Maximum	8.88	6.8	6.64	10.16	8.96	7.92
Maximum Slope	1.15	1.12	0.62	0.96	0.72	0.62
SVD Values	56.35	47.104	51.2794	103.2297	97.6427	49.415
Labeling	Normal	Normal	Abnormal	Abnormal	Abnormal	Normal

The Table 1 shows the feature extracted values along with the labels done by SVD for just six segments. As the PPG data sample length obtained is 1, 44, 000 and it is segmented into equal intervals comprising of 200 samples totally with the entire data consisting of 720 segments, here in table 1, it shows only 6 segments for understanding purpose. For the rest of the 714 segments, similar types of 10 features are extracted. From table 1, it is obvious that each segment has 20 samples and 10 features are extracted from it. Finally the feature extracted annotated values of PPG are classified and validated with the help of a GMM classifier.

4 GMM Classifier

A Bayesian model based GMM Classifier is used for classification here. To assess the class posterior probability, the Bayesian rule is followed as

$$q(d|h^n) = \frac{q(h^n|d)q(d)}{\Sigma_{c'=1}^{C} q(h^n|d')q(d')}$$

where $d \in \{1, ..., D\}$ and D represent the total number of classes. The denominator can be easily neglected as it only scales $q(d|h^n)$. The posterior probability $q(d|h^n)$ is used to model the probability of d, if the feature vectors of the nth sample h^n is given. Using Maximum posterior (MAP) estimation, the class label d^* is estimated, i.e, making optimum use of the class posteriors the class label d is determined as

$$d^* = \arg\max_{d\in\{1,....,D\}} q(d|h^n) = \arg\max_{d\in\{1,....,D\}} q(h^n|d)q(d)$$

The above equation is a solution of the minimization problem involving the Bayesian Risks with a 0/1-loss function. The class prior distribution is represented by the term $q(d)$. GMM is used for modeling the terms $q(h^n|d)$, i.e., for each and every class d , a GMM $q(h^n|\Theta_d)$ is produced. A GMM $q(h^n|\Theta d)$ is actually the weighted sum of $F > 1$ Gaussian components $N(h^n|\theta_d^f)$ in R^g, $q(h^n|\Theta_d)=\sum_{f=1}^{F}\alpha_d^f N(h^n|\theta_d^f)$,, where α_d^f signifies the weight of each component f = {1,...., F}. These weights are assumed or constrained to be positive values as $\alpha_d^f \geq 0$ and $\Sigma_{f=1}^{F}\alpha_d^f = 1$. The GMM is represented by a set of specific parameters, $\Theta_d = \{\alpha_d^1, \alpha_d^2, ..., \alpha_d^f, \theta_d^1, \theta_d^2, ..., \theta_d^F\}$, , where the mean vector μ_d^f specifies the Gaussians and the covariance matrix is specified by Σ_d^f. i.e, $\theta_d^f = \{\mu_d^f, \Sigma_d^f\}$. The most commonly used algorithm which is used alternately with the $\log q(J_d|\Theta_d) = \log\prod_{n=1}^{N_d} q(h^n|\Theta_d)$ converges to a local optimum point [22]. Only depending on the choice of the initial parameters, the performance of the EM algorithm is determined.

5 Results and Discussion

For the Performance Analysis of GMM Classifier for Classification of Normal and Abnormal Segments in PPG Signals, the bench mark parameters taken for the analysis is Sensitivity, Specificity, Precision, False Discovery Rate, Accuracy, Fl score and the values are computed in Table d. The sensitivity, specificity, precision, FDR, accuracy and Fl score is expressed by the following mathematical formulae as

$$Sensitivity = \frac{TP}{TP+FN}$$

$$Specificity = \frac{TN}{TN+FP}$$

$$Precision = \frac{TP}{TP+FP}$$

$$FDR = \frac{FP}{TP+FP}$$

$$Accuracy = \frac{TP+TN}{TP+FP+FN+TN}$$

$$F1\ Score = \frac{2TP}{2TP+FP+FN}$$

Table 2 Consolidated Analysis of a GMM Classifier

PARAMETERS	Values
Sensitivity (%)	97.95%
Specificity (%)	100%
Precision (%)	100%
FDR (%)	0
Accuracy (%)	98.97%
FI Score	98.96%

6 Conclusions

PPG Signals are used for various purposes such as to monitor heart rate and cardiac cycle, respiration, depth of anesthesia, hypovolemia, hypervolemia etc. Therefore the volumetric measurements of a particular organ can be provided by a PPG. With the help of a pulse oximeter, PPG readings are obtained where the skin is illuminated and the changes in the light absorption is measured. Here in this paper, the PPG reading of a single patient who is suffering from respiratory disorder is analyzed. The features were extracted successfully from the PPG signals of the patient and then classified with GMM Classifier. Results show that an accuracy of 98.97% is obtained, a sensitivity of about 97.95% is obtained, a specificity of about 100%, nil FDR and FI score of 98.96% is obtained. Future works may analyze the case for a lot of patients employing various other types of post classifiers for the classification of the extracted features to identify the normal and abnormal PPG segments for a patient with respiratory disorder.

References

[1] Bhattacharya J, Kanjilal P P and Muralidhar V Analysis and characterization of photo-plethysmographic signal *IEEE BME* **48** 5-112001.

[2] Gil E etal Heart Rate Turbulence Analysis Based on Photoplethysmography IEEE Transactions on Biomedical Engineering, **60**,2013

[3] Azabji Kenfack M, Lador F, Licker M, Moia C, Tam E, Capelli C, Morel D and Ferretti G Cardiac output by Modelflow method from intra-arterial and fingertip pulse pressure profiles *Clin. Sci.* **106** 365-9,2004

[4] J. A. Sukor, *et al.*, "Signal quality measures for pulse oximetry through waveform morphology analysis," *Physiological Measurement,* vol. 32, p. 369, 2011.

[5] J. Foo and S. Wilson, "A computational system to optimise noise rejection in photoplethysmography signals during motion or poor perfusion states," Medical and Biological Engineering and Computing, vol. 44, pp. 140-145, 2006.

[6] J. M. Graybeal and M. T. Petterson, "Adaptive filtering and alternative calculations revolutionizes pulse oximetry sensitivity and specificity during motion and low perfusion," in *Engineering in Medicine and Biology Society, 2004. IEMBS '04. 26th Annual International Conference of the IEEE*, 2004, pp. 5363-5366.

[7] Y.-s. Yan, *et al.*, "Reduction of motion artifact in pulse oximetry by smoothed pseudo Wigner-Ville distribution," *Journal of NeuroEngineering and Rehabilitation,* vol. 2, p. 3, 2005

[8] B. S. Kim and S. K. Yoo, "Motion artifact reduction in photoplethysmography using independent component analysis," *Biomedical Engineering, IEEE Transactions on,* vol. 53, pp. 566-568, 2006.

[9] Huotari, M., Yliaska, N., Lantto, V., Maatta, K., Kostamovaara, J. (2009). Aortic and arterial stiffness determination by photoplethysmographic technique. *Procedia Chemistry,* 1 (1), 1243-1246.

[10] Gavish, B. (1987). Photoplethysmographic characterization of the vascular wall by a new parameter- minimum rise-time: Age dependence on health. *Microcirc Endothelium Lymphatics,* 3 (3-4), 281-296.

[11] Reisner, A., Shaltis, P.A., McCombie, D., Asada, H.H. (2008). Utility of the photoplethysmogram in circulatory monitoring. *Anesthesiology*, 108 (5), 950-958.

[12] Mohamed, S., Mahamod, I., Zainol, R. (2004). Artificial neural network (ANN) approach to PPG signal classification. *International Journal of Computing & Information Sciences*, 2 (1), 58-65.

[13] Edmond, Z., Kalaivani, C., Mohd, A., Harwant, S.(2007). Analysis of the effect of ageing on rising edge characteristics of the photoplethysmogram using a modified Windkessel model. *Cardiovascular Engineering,* 7 (4), 172-181.

[14] Rubins, U., Grube, J., Kukulis, I. (2008). Photoplethysmography analysis of artery properties in patients with cardiovascular diseases. In *14th Nordic- Baltic Conference on Biomedical Engineering and Medical Physics,* 16-20 June 2008. Springer, 319 322.

[15] Qawqzeh, Y., Ali, M.A.M., Reaz, M.B.I., Maskon, O. (2010). Photoplethysmogram peaks analysis in patients presenting with erectile dysfunction. In *Proceedings of the International Conference on Electrical Computer Technology (ICECT2010),* 7-10 May 2010, 165-168.

[16] Loukogeorgakis, S., Dawson, R., Phillips, N., Martyn C.N., Greenwald, S.E. (2002). Validation of a device to measure arterial pulse wave velocity by a photoplethysmographic method. *Physiological Measurement,* 23 (3), 581596.

[17] Gonzalez, R., Delgado, A., Padilla, J., Trenor, B. Ferrero, J., Saiz J. (2008). Photoplethysmographic augmentation index as a non invasive indicator for vascular assessments. In *4th European Conference on the International Federation for Medical and Biological Engineering,* 23-27 November 2008. Springer, 1167-1170.

[18] M. Saeed, M. Villarroel, A.T. Reisner, G. Clifford, L. Lehman, G.B. Moody, T. Heldt, T.H. Kyaw, B.E. Moody, R.G. Mark. Multiparameter intelligent monitoring in intensive care II (MIMIC-II): A public-access ICU database. Critical Care Medicine 39(5):952-960 (2011 May); doi: 10.1097/CCM.0b013e31820a92c6.

[19] Goldberger AL, Amaral LAN, Glass L, Hausdorff JM, Ivanov PCh, Mark RG, Mietus JE, Moody GB, Peng C-K, Stanley HE. PhysioBank, PhysioToolkit, and PhysioNet: Components of a New Research Resource for Complex Physiologic Signals. *Circulation* **101**(23):e215-e220 [Circulation Electronic Pages; http://circ.ahajournals.org/cgi/content/full/101/23/e215]; 2000 (June 13)

[20] Sunil Kumar Prabhakar, Harikumar Rajaguru, "Morphological Operator Based Feature Extraction Technique along with Suitable Post Classifiers for Epilepsy Risk Level", Proceedings of the International Conference on Intelligent Informatics and BioMedical Sciences (ICIIBMS), November 2830, Okinawa, Japan

[21] Harikumar Rajaguru, Sunil Kumar Prabhakar, " An Exhaustive Analysis of Code Converters as Pre-Classifiers and K means, SVD, PCA, EM, MEM, PSO, HPSO and MRE as Post Classifiers for Classification of Epilepsy from EEG Signals", Journal of Chemical and Pharmaceutical Sciences, Vol.9(2), pp:818-822, April-June 2016

[22] Sunil Kumar Prabhakar, Harikumar Rajaguru, " ICA, LGE and FMI as Dimensionality Reduction Techniques followed by GMM as Post Classifier for the Classification of Epilepsy Risk Levels from EEG Signals", 9th IEEE European Modelling Symposium 2015, October 6-8, Madrid, Spain, 978-15090-0206-1/15, DOI: 10.1109/EMS.2015.20

Development of Patient Remote Monitoring System for Epilepsy Classification

Sunil Kumar Prabhakar, Harikumar Rajaguru

Department of ECE, Bannari Amman Institute of Technology, India

Abstract. Epilepsy is one of the most commonly occurring neurological disorders and is characterized by hyper-synchronous neuronal firing. The unexpected and continous electrical disturbances in the brain can be very disturbing to the patient thereby affecting the overall quality and happiness of the patient's life. Electroencephalography (EEG) signals helps in the analysis and diagnosis of epilepsy by recording the activities in the cortical regions of the epileptic patient. Epilepsy is characterized by the abnormal EEG signal flow. The recordings of the EEG signals are too huge to process and hence the necessity of dimensionality reduction is mandatory. The total number of epileptic patients analyzed in this work is twenty. Initially the dimensions of the EEG signal are reduced with the help of Fuzzy Mutual Information (FMI). The dimensionally reduced data is then transmitted with the help of a 2 x 2 Orthogonal Space Time Block Coded Multiple Input Multiple OutputOrthogonal Frequency Division Multiplexing (OSTBC MIMO-OFDM) System. As the OSTBC MIMO-OFDM System suffers a high Peak to Average Power Ratio (PAPR), a unique approach of Singular Value Decomposition Based Partial Transmit Scheme (SVD-PTS) is proposed to reduce the PAPR and Bit Error Rate (BER) at the receiver side. Also at the receiver side, Gaussian Kernel Based Support Vector Machine (G-SVM) with a Kernel value of 10 is utilized thoroughly in this work. The performance metrics such as Specificity, Sensitivity, Time Delay, Quality Value, Performance Index and Accuracy are analyzed here. Results show that an average accuracy of 95.38% is obtained, average perfect classification rate is 90.76%, time delay is 2.19 seconds and quality value is 20.53.

Keywords: Epilepsy, EEG, FMI, SVD-PTS, Accuracy

1 Introduction

To control the overall coordination of the nerves and muscles in the human body, brain is utilized and it is considered as one of the complex parts of the entire human body. There are a lot of neurological disorders which affect the brain and one such disorder is epilepsy [1]. Considered as one of the most commonly occurring neurological disorder, epilepsy is a huge threat to the mankind. It is often recognized as a chronic condition of the human nervous system. Stemming from sudden and temporary abnormal discharges of the electrical activities of the brain, epilepsy occurs and it leads to unprovoked seizures. It is also known as fits because it causes loss of consciousness and memory, various alterations in behaviour, emotion and sensation. All the epileptic patients cannot be cured with the help of medicines and antibiotics, because for some patients medicines may not respond and cause a lot of allergy. So depending on antiepileptic drugs is not the only solution for such people.

To identify the psychological and physiological situations in the human and to trace the functional activities of the brain, EEG is used [2]. As EEG is highly non linear and non-stationary, a good analysis is required to differentiate between normal EEG and seizure affected EEG signals. Thus analyzing the signals of the brain through EEG lays a direct form of communication between the physician and the brain. EEG serves as a good clinical tool for the assessment of human activity and it gives valuable data of the patients who suffer from epilepsy. For the epileptic patient, the EEG signal flow is quite abnormal and the amplitude of the spikes generated in the EEG waveform need not always represent the seizure severity and it is important to differentiate between the seizure period and the period between the seizure stages [3]. So to record the electrical activities of the brain, EEG seems to be the most flexible, easy to use technique and has a relatively lower cost. Since the recordings of the EEG are long, the detailed analysis of it seems to be very difficult as it requires trained and specialized visual encephalographers. Moreover the decision can be erroneous due to a lot of factors. So in the past two decades a lot of reliable automated detection algorithms for the classification and detection of epilepsy has been proposed. Since this work has been implemented for telemedicine applications, some of the top works done in the field of seizure classification and the implementation of it to the wireless telemedicine applications is discussed here. A general framework to assess and compare seizure prediction methods was given

J. Goh et al. (eds.), *The 16th International Conference on Biomedical Engineering*,
IFMBE Proceedings 61,
DOI: 10.1007/978-981-10-4220-1_16

by Winterhalder et al [4]. The seizure anticipation by patients with focal and generalized epilepsy and a multicentre assessment of premonitory symptoms was given by Schulze et al [5]. The epileptic seizure prediction using hybrid feature selection over multiple EEG electrode contacts was reported by D'Alessandro et al [6]. The time delay neural network and ICA for EEG based prediction of epileptic seizures propagation was reported by Mirowski et al [7]. The recurrent neural network based prediction of epileptic seizures in intra and extracranial EEG done by Petrosian et al [8]. Testing statistical significance of multivariate time series analysis techniques for epileptic seizure prediction was done by Schelter et al [9]. The Factorization concept and Particle Swarm Based Sparse Representation Classifier for Epilepsy classification was implemented for Wireless Telemedicine Applications was developed by Sunil and Harikumar [10]. The epilepsy diagnosis using probability density function of EEG signals was reported by Orhan et al [11]. The Epilepsy Classification Using Discriminant Analysis and Implementation with Space Time Trellis Coded MIMO- OFDM System for Telemedicine Applications was modelled by Sunil and Harikumar [12]. The automatic epileptic seizure detection in EEGs based on line length feature and ANN was reported by Guo et al [13]. A robust principal component analysis algorithm for EEG-based vigilance estimation is done by Shi et al [14]. The Hadamard Transform Based PAPR Reduction for Telemedicine Applications Utilized for Epilepsy Classification by Sunil and Harikumar [15]. A PCA Based Selective Mapping Technique for Reduced PAPR Implemented for Distributed Wireless Patient Monitoring Epilepsy Classification System was developed by Sunil and Harikumar [16]. Entropy Based PAPR Reduction for STTC System Utilized for Classification of Epilepsy from EEG Signals Using PSD and SVM was modelled by Sunil and Harikumar [17]. Wireless Systems with Reduced PAPR Using K-means Modified PTS Implemented for Epilepsy Classification from EEG Signals was developed by Sunil and Harikumar [18]. The Efficient Wireless System for Telemedicine Application with Reduced PAPR Using QMF Based PTS Technique for Epilepsy Classification from EEG Signals was modelled by Sunil and Harikumar [19].

In this work, using FMI, the dimensions are reduced and then it is transmitted through a 2 x 2 OSTBC MIMO-OFDM system. Since the system suffers from a high PAPR, SVD Based PTS (SVD-PTS) is used to reduce the PAPR and at the receiver side, GK-SVM is used to classify the epilepsy from EEG signals. Figure 1 shows the block diagram of the system.

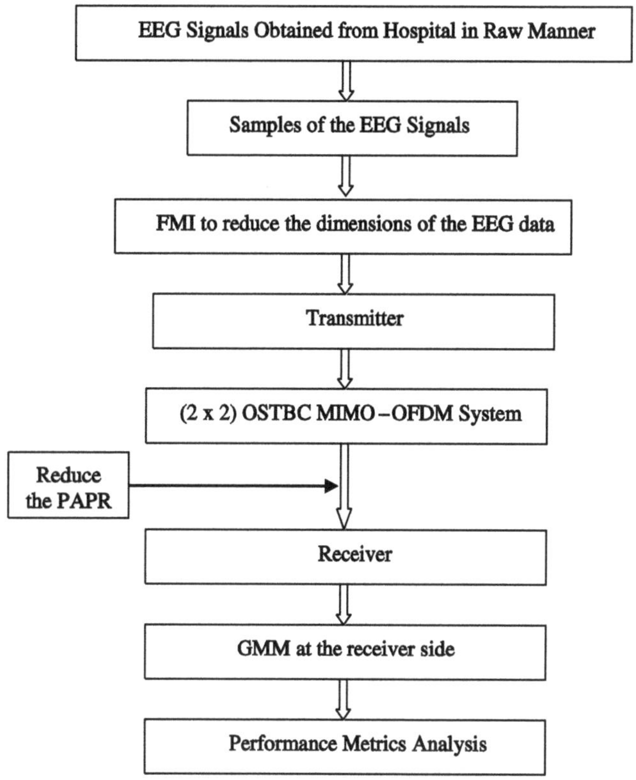

Figure 1 Block Diagram of the Entire Work

2 Materials and Methods

In this section, the acquisition of EEG data, the need for dimensionality reduction and how FMI is employed to reduce the dimensions of the EEG data is discussed here.

2.1 Acquisition of EEG Data and the Preference for Dimensionality Reduction

From the Department of Neurology of Sri Ramakrishna Hospital, Coimbatore, India, readings of 20 epileptic patients who were taking their treatment in the hospital was taken for this evaluation. The required permission was obtained from both the patients and doctors. The obtained EEG data set is in European Data Format (EDF). According to the International 10-20 system, the 16 channel electrodes are placed on the scalp of the patient. Before the placement of the electrode on the scalp, it is cleaned well using any cleansing agent and then with the help of electrode paste, it is placed on to the scalp of the patient. The referencing of the electrodes is done to a common potential point like ear. For different periods like awake stage, sleeping stage, resting stage and so on the readings are taken at a speed of 30 mm/s for more than 40 minutes duration. Using suitable EEG machines, amplification of the signals can be done easily. As there are 16 channels for each patient, each channel recording is split into three or four epochs with two second duration. The two second epoch is widely preferred because it can easily detect and trace the most vital changes happening in the signal. For this analysis the tool employed to do the work is MATLAB 7.0 version. Since 50 Hz is the frequency of the EEG signal, the sampling frequency has to be twice more than it based on the Nyquist criterion and sampling theorem and so 200 Hz is taken as the sampling frequency. Each and every sample of the data will have the instantaneous amplitude values of the signal and so totally 400 sample values are present for a single epoch in a single channel. So to process all the sample values for the epochs in all the channels for the entire twenty patients is a very difficult job to perform and so dimensionality reduction is required.

2.2 Dimensionality Reduction Using Fuzzy Mutual Information

Mutual information in general tells how much data one random variable conveys about another random variable. The Mutual Information for any two random variables a and b denoted by $I(a,b)$ is expressed as

$$I = \sum_{v_a \in dom(a)} \sum_{v_b \in dom(b)} p(v_a, v_b) \log \frac{p(v_a, v_b)}{p(v_a) p(v_b)}$$

The information which b tells about a is just the reduction measures in uncertainity about a due to the in depth knowledge of b and vice-versa. If the value of $I(a,b)$ is more, then a lot of information is told by both a and b . The normalized mutual information for the two attributes a and b, represented as $\tilde{I}(a,b)$ is expressed as $\tilde{I}(a,b) = \frac{I(a,b)}{I(a,a)}$, in such a manner that $I(a,a) = H(a)$ and the rules are then constructed. For FMI, the steps are as follows:

a) For the qualitative data, the pre-processing step is done which includes defining fuzzy sets and membership function
b) Then the Normalized MI is computed for each and every pair of attributes.
c) Using NMI, the potential and frequent item sets are selected
d) The association rules are then generated with respect to the attributes selected alone with the fuzzy sets
e) The adjusted difference for all the rules generated is calculated.
f) Based on the adjusted difference, the filtering rules can be modeled.
g) The filtered rules have to be joined for generating more rules.
h) Unless a new rule is generated, the process is repeated. The fuzzy association rules designed will reduce the dimensions of the data.

The preprocessing using Fuzzy K means clustering is done initially. For finding the membership of each value, fuzzy k-means clustering is used. For the clustering of the attribute value, the total number of 'k' points is placed into the space represented by the objects which are prone to clustering. Such points represent the initial group centroid. Each object is assigned to the group which has the closest centroid. After the assignment of all objects, the position of the k- centroid is recalculated. Until the centroid can move no longer, the process is continued. The main aim is to reduce the objective function (squared error function)

$$J = \sum_{i=1}^{N} \sum_{j=1}^{k} (u_{ij})^m (a_i - c_j)^2$$

where the distance measure chosen between the attribute value $a_i^{(j)}$ and the cluster centre c_j. is represented as $(a_i - C_j)^2$. It indicates the distance of the number of records (N). The resulting clusters are associated with the 'k' linguistic terms. Based on the cluster centres and the attribute nature, the linguistic terms are associated and finally the membership value is computed for each and every value. The Normalized Mutual information is computed. Each quantitative value is transformed into a fuzzy set which has linguistic terms with the help of membership functions and then the normalized mutual information is calculated for all attributes on the transaction data. In between the distinct pair of attributes, the value of normalized mutual information is calculated [20]. If the two adjusted variables are denoted as a_i and b_j , then $\tilde{I}(a_i,b_j) \geq \mu$ represents a very strong relationship in between the attributes. The rules are generated and then selection of the rules is done accordingly. The interesting association between the two attribute values is identified and as a result the dimensionality is reduced.

3 Design for Telemedicine Application

As the epilepsy classification is designed for Telemedicine Application, the dimensionally reduced values are transmitted through the system implemented here, that it, Orthogonal Space Time Block Coded MIMO-OFDM System. As the system has a high PAPR, SVD-PTS algorithm is used to reduce the PAPR.

3.1 OSBTC MIMO-OFDM System

In the past decade, OFDM has attracted a lot of interest for the standardization purposes of wireless systems. OFDM technique is a multicarrier modulation technique implemented with the help of FFT/IFFT algorithms. It is highly robust against frequency-selective fading channels and so it is adopted for transmission system such as Worldwide interoperability for Microwave Access (WiMAX), Wireless Fidelity (Wifi), Digital Video Broadcasting (DVB) and Long Term Evolution (LTE). To improve the spectral efficiency and to enhance the link reliability, OFDM is combined with MIMO [21]. To compensate the problem of fading, MIMO is used. In MIMO, the same signals can be sent through the different MIMO antennae and the different antennae at the receiver side will receive the signals travelled from diverse paths. For increasing the capacity of the channel MIMO is used such that different set of data is transmitted at the same time through different MIMO antennae without the automatic-repeat request of the transmission. In MIMO system, to attain maximum efficiency, the spacing of the antennas needs to be at least half of the transmitted signal. FFT algorithm is used for the efficient implementation of MIMO-OFDM system enhanced by MIMO encoding techniques such as Space Time Block Codes and Space Time Trellis Codes (STTC). In this paper, Orthogonal Space Time Block Codes (OSTBC) is used [22]. Space time block coding is a great boon for communication over the Rayleigh channels with the help of multiple transmits antennas. Using a space-time block code, data is encoded and it is split into $'n'$ streams. The $'n'$ streams are transmitted simultaneously by using $'n'$ transmit antennas. The signals which are received at the receiver antenna are nothing but a linear superposition of the total $'n'$ transmitted signals which is perturbed by noise. Through the decoupling operation of the signals which are transmitted from various antennas, maximum likelihood decoding is achieved. This makes use of the orthogonal structure of the STBC and so a maximum-likelihood decoder by linear processing is operated at the receiver. These codes are designed for the maximum achievement of the diversity order for a total number of transmit and receive antennas. It is simply subject to the condition of having a simple decoding algorithm.

3.2 PAPR Problem in OSTBC MIMO-OFDM and SVD-PTS technique to mitigate it:

As the OSTBC MIMO-OFDM system is engaged here, a high PAPR is a problem and using a modified algorithm called SVD Based PTS, the PAPR is reduced to a great extent.

Some of the peak values of the transmitted signals are larger than the typical values. If the PAPR is high in the transmitted signals of the OFDM, then there is a reduction in BER performance, intermodulation effects on the sub carriers, the energy is spilled into the adjacent channels and the power amplifiers suffer from non linear distortion [23]. The reason for a high PAPR in the time domain of the OFDM signal is due to the actual sum of the narrowband signals. At some places, the total sum of those narrowband signals is much larger than the average values. So with this high PAPR, if the signals are transmitted through a non linear power amplifier, a spectral broadening is created and the dynamic range of the Digital to Analog Converter is increased. As a result, the cost of the system increases and the efficiency reduces. The time domain representation of the complex baseband OFDM signal is expressed as

$$z_n = \frac{1}{\sqrt{N}} \sum_{k=0}^{N-1} Z_k e^{j2\pi nk/N}, n = 0,1,2,\ldots,N-1$$

where z_n represents the nth component in the OFDM output symbol, Z_k represents the kth modulated data symbol in the frequency domain of OFDM, N represents the number of subcarriers.

The PAPR (dB) of the transmitted OFDM signal is represented as

$$PAPR = \frac{\left(\max|z_n|^2\right)}{E\left[|z_n|^2\right]}$$

where $E[.]$ denotes the expected value operation. The theoretical maximum of the PAPR for Z number of sub carriers is as follows

$$PAPR_{max} = 10\log(N)\ dB$$

PAPR is represented as a random variable as it is a function of input data. The Complementary Cumulative Distribution Function (CCDF) for various PAPR values is given as

$$CCDF = P_r(PAPR > PAPR_0)$$

The SVD Based PTS is implemented as follows to get a reduced PAPR

1) The inputs are generated randomly from the input data block.
2) It is multiplied by various phase sequences.
3) A modified data block is obtained after multiplying it with various phase sequences
4) To get the time domain series, the independent sequences are inserted into IFFT
5) The choosing of the side information is done so that the PAPR is minimized.

6) After this, the Side Information (SI) is multiplied with the mapped input data.
7) The symbols with the least PAPR are disjoined into sub blocks.
8) For each of the partitioned sub blocks, SVD is applied and then multiplied by a corresponding complex phase factor [24].
9) The least PAPR is obtained by choosing the preferred phase vector.
10) All the sets of phase vectors are searched to find the optimum set of phase factor.

The simulation parameters of the system designed are shown in Table 1.

Table 1 Simulation Parameters

Modulation Technique Used	QPSK
System analysed	2 x2 OSTBC MIMO-OFDM
Nature of Space Time Block Codes	Orthogonal
Number of subcarriers	512
No of sub-blocks	4
Maximum symbols loaded	1e5
Symbol rate Analyzed	250000
Total No. of time slots	2
Windowing function used	Blackman-Haris
HPA Model analysed	SSPA
No of frames utilized	10
Total No of OFDM symbols/ frame	4
Bandwidth Taken	5 MHz
Oversampling factor Used	4
No of Phase Factors in PTS Scheme	4

4 Support Vector Machine (Gaussian Kernel) at the Receiver

The SVM is a machine learning algorithm which is utilized to solve classification problems [25]. It utilizes a flexible representation of the boundary classes. To reduce over fitting, it implements the automatic complexity control. It has a single global minimum value which is found in the polynomial time. It is highly popular because, it can be easily used, and it has a very good generalization performance. With a little tuning, the same algorithms are used to solve a variety of problems.

The main goal of it is to separate the two classes without the loss of generality. The separation of the set of training vector which belongs to two different classes are represented as

$$E = \{(a^{'},b^{'}).....(a^{l},b^{l})\}, a \in R^{n}, b \in \{-1,1\}$$

with a respective hyperplane represented as

$$\langle w,a\rangle + c = 0$$

It is said to be optimally represented by the hyperplane if the separation is done without errors. Also the distance between the hyperplane to the closest vector should be maximal for the set of vector to be optimally separated. Without the loss of generality, it is good to consider a canonical hyperplane, where the parameters w, c are constructed by

$$\min_{i} \left|\langle w,a^{i}\rangle + c\right| = 1$$

The constraint should be preferable to the simplification of the formulation of problem. The norm of the weight vector should be definitely equal to the inverse of the distance points, closest to the hyperplane.

A canonical form, a separating hyperplane must meet the following constraint as

$$b^{i}\left[\langle w,a^{i}\rangle + c\right] \geq 1, i = 1,...l$$

The distance is represented as $d(w,c;a) = \frac{\left|\langle w,a^{i}\rangle + c\right|}{\|w\|}$

By maximizing the margin, the optimal hyperplane is obtained. For the optimal separation of data, a hyperplane is used so that it minimizes the condition

$$\phi(w) = \frac{1}{2}\|w\|^{2}$$

With the explanation of the technique to obtain an optimal hyperplane, the support vector machine classifier for a function $f(q)$ with Gaussian Kernel is represented as $f(q) = \sum_{i}^{N} \alpha_i b_i k(q_i, q) + c$, where α_i are known as weights and can be zero too sometimes, q_i is denoted as the support vector here. The Gaussian kernel representation $k(q,q') = \exp\left(-\|q-q'\|^{2} / 2\sigma^{2}\right)$. The Kernel value is set to be around 10 in this case.

5 Results and Discussion

For the telemedicine application implemented for the STBC MIMO-OFDM System engaging FMI as a dimensionality reduction technique and Gaussian Kernel Support Vector Machine (SVM) as a post classifier at the receiver side, the results are analyzed and computed in this section

5.1 PAPR Results:

From the Figure 2, it is evident that a PAPR reduction of about 3.5 dB is achieved if the SVD-PTS algorithm is employed for the OSTBC MIMO-OFDM System. The Bit Error Rate is tabulated in Table 2.

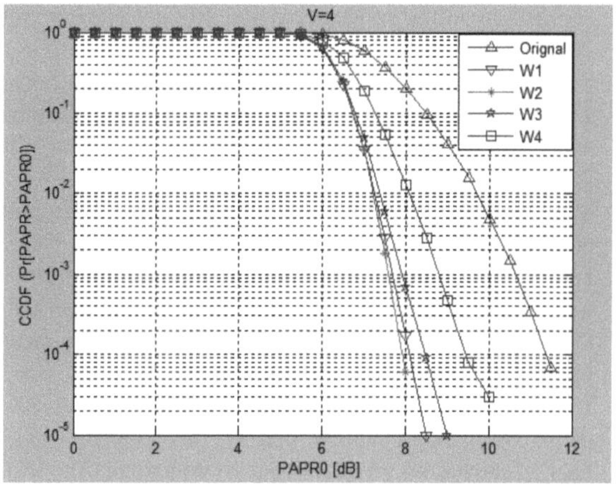

Figure 2 PAPR Result Analysis usign SVD-PTS Algorithm

Table 2 BER Analysis

SNR	BER
0	0.1416
2	0.1218
4	0.1109
6	0.0913
8	0.0723
10	0.0514
12	0.0218
14	0.0048

5.2 Classification Results:

The standard performance metrics considered here are Performance Index (PI), Sensitivity, Specificity and Accuracy are tabulated in Table 3 and are given by the respective formulae as follows

$$PI = \left(\frac{PC - MC - FA}{PC} \right) \times 100$$

where PC stands for Perfect Classification, MC stands for Missed Classification and FA stands for the False Alarm. The Sensitivity, Specificity and Accuracy measures are stated by the following

$$Sensitivity = \frac{PC}{PC + FA} \times 100$$

$$Specificity = \frac{PC}{PC + MC} \times 100$$

$$Accuracy = \frac{Sensitivity + Specificity}{2} \times 100$$

The Quality Value Qv is defined as

$$Q_v = \frac{C}{(R_{fa} + 0.2) * (T_{dly} * P_{dct} + 6 * P_{msd})}$$

C

where the scaling constant is represented by ***C***

The total number of false alarm per set is represented as R_{fa}

The average time delay of the onset classification (seconds) is represented as T_{dly}

The percentage of perfect classification is given as P_{dct} and

The percentage of perfect risk level missed is denoted by P_{msd}

The time delay is given as follows

$$Time\ \ Delay = \left[2 \times \frac{PC}{100} + 6 \times \frac{MC}{100} \right]$$

Table 3 Average Classification Results of FMI-GKSVM at the Receiver Side

PARAMETERS	Epoch -1	Epoch -2	Epoch -3	Average
PC (%)	90.208	90.833	91.25	90.763
MC (%)	8.333	7.083	3.75	6.388
FA (%)	1.458	2.083	5	2.847
PI (%)	88.366	89.692	90.033	89.364
Sensitivity (%)	98.541	97.916	95	97.152
Specificity (%)	91.666	92.916	96.25	93.611
Time Delay (Sec)	2.304	2.241	2.05	2.198
Quality Value	20.748	20.694	20.157	20.533
Accuracy (%)	95.104	95.416	95.625	95.381

6 Conclusion

Thus in this paper, the dimensions of the EEG signal are reduced with the help of FMI and then transmitted through the OSTBC MIMO-OFDM System and then classified at the receiver side with the help of Gaussian Kernel SVM. The results were analyzed here in terms of PAPR, BER, PI, Specificity, Sensitivity, Accuracy, Time Delay and Quality Values. The results show that when SVD-PTS algorithm was used to reduce the PAPR in the system designed for telemedicine applications, the PAPR was reduced by 3.5 dB and a low BER was also achieved easily. If the classification results are analyzed in the receiver side, it shows that an average perfect classification as of 90.76%, average PI of 89.36%, average time delay of 2.198 seconds, average quality values of about 20.533 and an average accuracy of 95.38% is obtained. Future works plan to implement various system analysis designed for telemedicine applications, various post classification schemes and different dimensionality reduction schemes.

References

[1] Sunil Kumar Prabhakar, Harikumar Rajaguru, " Analysis of Centre Tendency Mode Chaotic Modeling for Electroencephalography Signals Obtained from an Epileptic Patient", Advanced Studies in Theoretical Physics, Vol.9,2015, no.4, 171-177, HIKARI Ltd, http:"dx.doi.org'10.12988'astp.2015. 5117.

[2] R.Harikumar, P.Sunil Kumar, "Frequency behaviors of electroencephalography signals in epileptic patients from a wavelet Thresholding perspective", Applied Mathematical Sciences, Vol. 9, 2015, no.50, 2451-2457, HIKARI Ltd, http:/'dx.doi.org'10.12988'ams.2015.52135.

[3] R.Harikumar, P.Sunil Kumar, "Dimensionality Reduction with Linear Graph Embedding Technique for Electroencephalography Signals of an Epileptic Patient", Research Journal of Pharmacy and Technology, Vol.8, No.5, pp: 554-556.

[4] WinterhalderM, Maiwald T, Voss HU, Aschenbrenner-Scheibe R, Timmer J. The seizure prediction characteristic: a general framework to assess and compare seizure prediction methods. Epilepsy Beh. 2003:4(3):318-325.

[5] Schulze-Bonhage A, Kurth C, Carius A, Steinhoff BJ, Mayer T. Seizure anticipation by patients with focal and generalized epilepsy: a multicentre assessment of premonitory symptoms. Epilepsy Res. 2006:70:83-88.

[6] D'Alessandro M, Esteller R, Vachtsevanos G, Hinson A, Echauz J, Litt B. Epileptic Seizure Prediction Using Hybrid Feature Selection Overn Mulitple EEG Electrode Contacts: A Report of Four Patients. IEEE Trans Biomed Eng. 2003:50(5):603-615.

[7] Mirowski P, Madhavan D, LeCun Y. Time-Delay Neural Networks and Independent Component Analysis for EEGBased Prediction of Epileptic Seizures Propagation. In: Proceedings of the Twenty-Second AAAI Conference on Artificial Intelligence, July 22-26, 2007, Vancouver, BC, Canada. AAAI Press, 2007:1982-1983

[8] Petrosian A, Prokhorov D, Homan R, Dasheiff R, Wunsh II D. Recurrent neural network based prediction of epileptic seizures in intra- and extracranial EEG. Neurocomputing. 2000:30:201218.

[9] Schelter B, Winterhalder M, Maiwald T, Brandt A, Schad A. Testing statistical significance of multivariate time series analysis techniques for epileptic seizure prediction. Chaos. 2006: 16: 013108.

[10] Sunil Kumar Prabhakar, Harikumar Rajaguru, "Factorization and Particle Swarm Based Sparse Representation Classifier for Epilepsy Classification Implemented for Wireless Telemedicine Applications", Proceedings of 6th International Conference on the Development of Biomedical Engineering, pp: 474-478, Ho Chi Minh City, Vietnam.

[11] U. Orhan, M. Hekim, M. Ozer, I. Provaznik, "Epilepsy diagnosis usingprobability density functions of EEG signals," Innovations in Intelligent Systems and Applications (INISTA), Istanbul, Turkey, 15-18 June 2011, pp. 626 - 630.

[12] Sunil Kumar Prabhakar, Harikumar Rajaguru, "Epilepsy Classification Using Discriminant Analysis and Implementation with Space Time Trellis Coded MIMO-OFDM System for Telemedicine Applications", Proceedings of 6th International Conference on the Development of Biomedical Engineering, Ho Chi Minh City, pp: 479-483, Ho Chi Minh City, Vietnam.

[13] L. Guo, D. Rivero, J. Dorado, J. R. Rabu~nal, and A. Pazos," Automatic epileptic seizure detection in EEGs based on line length feature and artificial neural networks," Journal of Neuroscience Methods 191, 2010, pp. 101 -109

[14] L. Shi, R. Duan, B. Lu, "A robust principal component analysis algorithm for EEG-based vigilance estimation," Engineering in Medicine and Biology Society (EMBC), 35th Annual International Conference of the IEEE, Osaka, Japan, 3-7 July 2013, pp. 6623 - 6626.

[15] Sunil Kumar Prabhakar, Harikumar Rajaguru, "Hadamard Transform Based PAPR Reduction for Telemedicine Applications Utilized for Epilepsy Classification", Proceedings of 3rd International Conference on Movement, Health and Exercise (MoHE), September 28-30, 2016, Malaysia.

[16] Sunil Kumar Prabhakar, Harikumar Rajaguru, "PCA Based Selective Mapping Technique for Reduced PAPR Implemented for Distributed Wireless Patient Monitoring Epilepsy Classification System", Proceedings of 3rd International Conference on Movement, Health and Exercise (MoHE), September 28-30, 2016, Malaysia.

[17] Sunil Kumar Prabhakar, Harikumar Rajaguru, "Entropy Based PAPR Reduction for STTC System Utilized for Classification of Epilepsy from EEG Signals Using PSD and SVM", Proceedings of 3rd International Conference on Movement, Health and Exercise (MoHE), September 28-30, 2016, Malaysia.

[18] Sunil Kumar Prabhakar, Harikumar Rajaguru, "Wireless Systems with Reduced PAPR Using K- means Modified PTS Implemented for Epilepsy Classification from EEG Signals", Proceedings of International Conference on Advancements of Medicine and Health Care through Technology (MEDITECH), October 12-15, 2016, Romania.

[19] Sunil Kumar Prabhakar, Harikumar Rajaguru, "Efficient Wireless System for Telemedicine Application with Reduced PAPR Using QMF Based PTS Technique for Epilepsy Classification from EEG Signals", Proceedings of International Conference on Advancements of Medicine and Health Care through Technology (MEDITECH), October 12-15, 2016, Romania.

[20] R.Harikumar, P.Sunil Kumar, "Fuzzy Mutual Information as a Dimensionality Reduction Technique for Epileptic Electroencephalography Signals ", Research Journal of Applied Science, Engineering and Technology, Maxwell Scientific Publishers, Vol.10, No.9, 1035-1037, 2015

[21] Seyran Khademi et al, " Constant Modulus Algorithm for Peak-to-Average Power Ratio (PAPR) Reduction in MIMO OFDM", IEEE Signal Processing Letters, Vol.20, No.5, May 2013

[22] H Yang (2005), "A Road to Future Broadband Wireless Access :MIMO-OFDM-Based Air Interface",IEEE Comm, Magazine, pp.53-60.

[23] Han S.H and Lee J.H (2005), " An Overview of Peak-to-Average Power Ratio Reduction Techniques for Multicarrier Transmission", IEEE Wireless Comm, pp.56-65.

[24] Harikumar Rajaguru, Sunil Kumar Prabhakar, "Assessment of Epilepsy Classification Using Techniques such as SVD, ApEn and Weighed KNN Measures", Asian Journal of Pharmaceutical and Clinical Research, Vol.9, Issue 5, pages:91-96, 2016

[25] Harikumar Rajaguru, Sunil Kumar Prabhakar (2016), "LDA, GA and SVM's for Classification of Epilepsy From EEG Signals", Research Journal of Pharmaceutical, Biological and Chemical Sciences, Vol.7(3), pp:2044-2049.

Discrete Wavelet Transform based statistical features for the Drowsiness detection from EEG

Reddy Vamsi[1], Dabbu Suman[2], CH Nikhil[1], M. Malini[2]

[1]Biomedical Department, National Institute of Technology, Rourkela, India.
[2] Dept. of BME, University College of Engineering, Osmania University, Hyderabad, India.

Abstract— Drowsiness is a major patron to road accidents. Detection of drowsiness while driving is a challenging objective in accidents avoidance systems. This study reports a new index to assess the drowsiness state of drivers using Joint Time-frequency analysis of Electroencephalography (EEG). Fifteen healthy male participants proffered in this study by performing a tedious driving task on a static simulator for 60 min. The subjects are deprived of sleep for at least 18 hrs and sleep music (Delta waves) played in the background, induces sleep during the task. Acquisition of EEG signals was implemented by eight channel Octal Bio-Amplifier (AD Instruments) at a sampling frequency of 1000Hz. The electrodes positioned at the four lobes of the brain namely Frontal, Temporal, Parietal, and Occipital lobes and further analysis has been carried out in MATLAB™ 2007b (Math works, Inc., USA). The EEG signals are de-noised by Chebyshev filter (0.5-40 Hz) and subsequently decomposed into various rhythms of EEG such as Beta (CD5: 14-30 Hz), Alpha (CD6: 8-13 Hz), Theta (CD7: 4-7 Hz), and Delta (CA7: 0.5-3.5 Hz). Two parameters viz., Relative Wavelet Packet Energy (RWPE) and Power within the RMSD (PRMSD) are computed in this study to analyse the driving performance of the subjects against the subjective assessment inferred from the video recordings. The parameters RWPE, PRMSD within Beta and Alpha has reduced relatively in the Parietal lobe, occipital lobe, and temporal lobe as the subjects fall into the drowsy stage. This analysis is clinically correlated as the cortical activity reduces slowly during the onset of sleep. It is clearly evident that these features are significant ($p < 0.05$) in the detection of drowsiness with Confidence Interval of 14. This study also reports a significant correlation ($p < 0.05$) between PRMSD of Total mean with Active mean and Drowsy mean with R^2 value of 0.82.

Keywords— Drowsiness, Relative Wavelet Packet Energy (RWPE), and Root Mean Square Deviation (RMSD)

1. INTRODUCTION

Universally, the macroscopic phase transition in the transportation made the roads utterly engaged with an ensemble of motorized vehicles. Road fatalities counts as the most probabilistic causes of death and gage 2.2% across the planet. Lamentably, drivers intermittently err the danger and adorn their adroitness to withstand the drowsiness behind the wheel. The National Highway Traffic Safety Administration declared that every year one lakh road accidents ensue owing to drowsy driving and driver fatigue [1][2]. Diseases such as Acute Insomnia and Sleep Apnea Syndrome can also induce drowsiness during prolonged driving. The National Transportation Safety Board (2010) has proclaimed that driver fatigue causes crashes which results in either bruises or ends up in death.

Several drowsiness detection technologies exist based on the fusion of various categorised parameters namely vehicle-based, physiological and behaviour parameters.Many researchers analysed the physiological study of EEG through different techniques namely Independent Component Analysis, Power Spectrum and Discrete Wavelet Packet Decomposition (DPWD). Belyavin et al., 1987 proposed that the decrease in Beta (β) activity is a valuable indicator to detect drowsiness [3]. Kecklund et al.,1993 reported the variation of energy levels amid listening to music. In spite of numerous attempts in drowsiness detection through Power within Spectrum (PSG) and Relative Wavelet Packet Energy (RWPE), the quantification of drowsiness states is yet uncertain.

This study aims at the quantification of drowsiness states of the subjects. This quantification is achieved through the assessment of Relative Wavelet Packet Energy (RWPE) and Power within RMSD (PRMSD). Some significant variations in physiological parameters were observed to detect drowsiness levels.

2. MATERIALS AND METHODS

2.1. Protocol and Simulation Task:

The experimental setup consists of a static driving simulator, Data Acquisition System, A Graphical User Interface (Lab Chart 8 software), Audio system, and a high definition camera as shown in fig.2.1. Fifteen male participants chosen are deprived of sleep for last 18 hours with a valid Light Motor Vehicle (LMV) license. The average height, weight, age of the participants is 1.72±0.08 m, 62.7±11.2 kg, and 27.76±10.77 years respectively. A predriving task of 30 mins is given before the experiment to familiarise about the

J. Goh et al. (eds.), *The 16th International Conference on Biomedical Engineering*,
IFMBE Proceedings 61,
DOI: 10.1007/978-981-10-4220-1_17

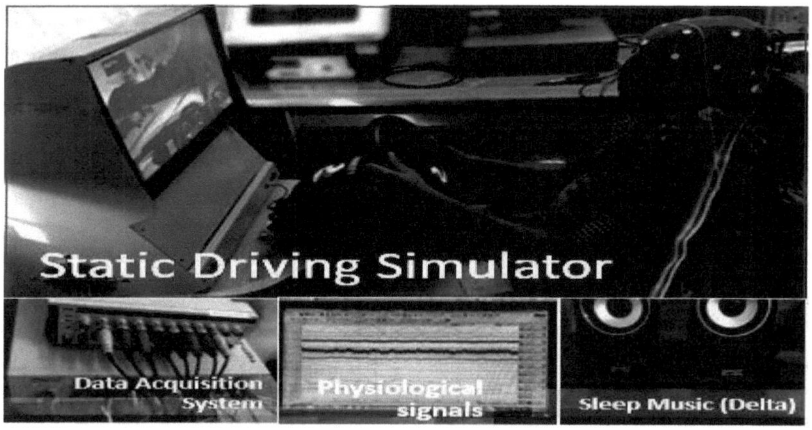

Fig.2.1. Experimental setup consists of a static driving simulator, Data Acquisition System

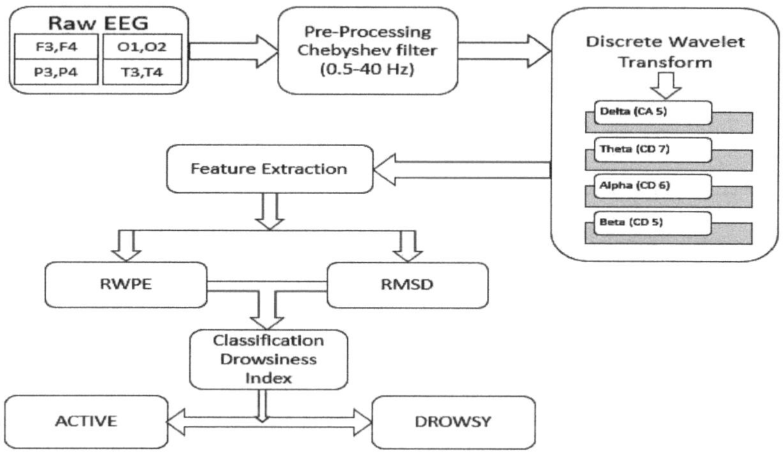

Fig.2.2. Physiological signal analysis

driving simulator and experimental protocol. A monotonous driving track is selected to perform the driving while sleep music (delta waves) is played in the background to induce the sleep. The participants performed the driving tasks on the static driving simulator for one hour. All the experiments are video graphed (16 MP: inbuilt-simulator) while driving to validate the results by subjective assessment. The calibration of the experimental setup, for the electrical safety by using Fluke- Electrical Safety Analyser (ESA115B) based on the standards IEC-60601-1 helps in assuring both patient and operator safety.

2.2 Physiological signal analysis:

The study uses multi-channel Octal Bio Amplifier from AD Instruments, Netherlands with Labchart8 software to acquire the EEG signals of the participants who are under the simulated driving condition. The steps followed in analysis adopted in this work are shown in Fig.2.2. The EEG signal acquired from the four lobes of the cerebral cortex, i.e., Frontal (F3, F4), parietal (P3, P4), occipital (O1, O2) and temporal (T3, T4) lobes based on 10-20 electrode system of EEG. EEG signals acquired from the subjects are analysed in the Mat Lab R 2007b to extract the features to classify the subjects into active and drowsy. A digital filter with cutoff frequencies 0.5-40 Hz is applied to remove the noise associated due to movement artifacts, eyeball movements, and power line interference. A non-stationary EEG signal is further processed by using Discrete wavelet transform (DWT).

Discrete Wavelet Transform (DWT) is a tool to convert the time domain signal into small waves (wavelets) with different frequency bands. DWT decomposition tree shown in Fig.2.3 consists of 2 filters High pass filter (Go), Low pass filter (Ho) at each stage, to divide the signal into different frequency bands, followed by the process of downsampling. This sort of sampling divides the length of the signal to exactly half at every stage to improve the

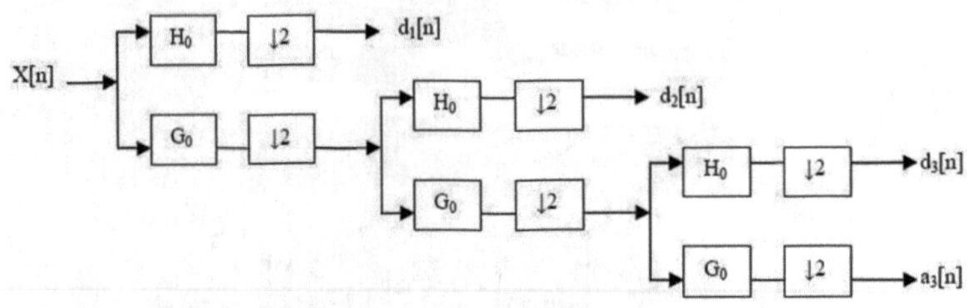

Fig.2.3. Decomposition tree of Discrete Wavelet Transform

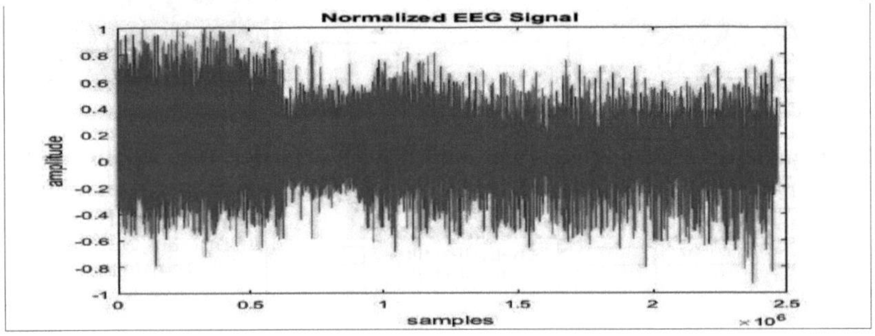

Fig.2.4. Normalized EEG signal

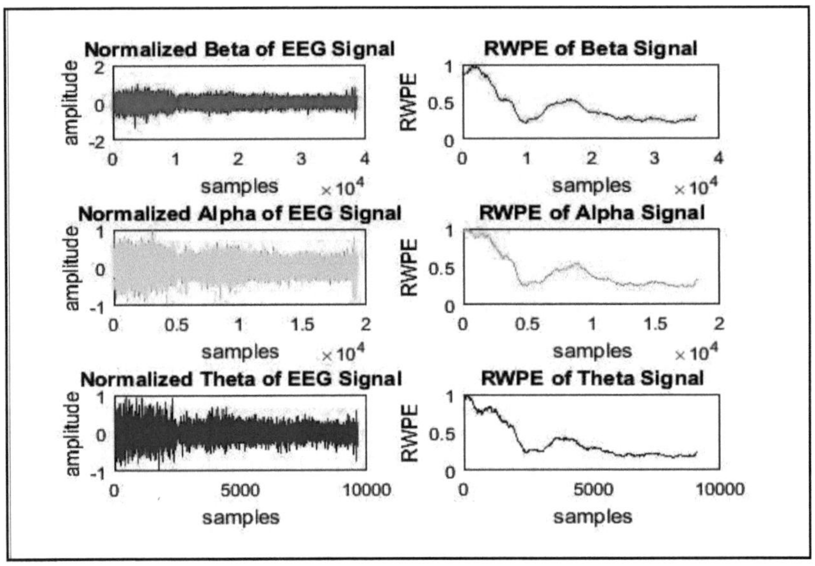

Fig.2.5. Decomposed EEG signals and their Relative Wavelet Packet Energy

frequency resolution. The DWT of the signal at every stage separates the signal into lowfrequency components a [n] called as the Approximation Coefficients (CA), high-frequency components d[n], called as the Detailed Coefficients respectively(CD).

The raw EEG signal acquired for nearly 1 hour with the sampling frequency of 1000 is shown in Fig.2.4, is decomposed up to 7 levels of the segmentation. The EEG data is divided into various rhythms of EEG such as Beta (CD5: 14-30 Hz), Alpha (CD6: 8-13 Hz), Theta (CD7: 4-7 Hz), and shown in Fig.2.5. The Feature Extraction of EEG rhythms Beta, Alpha, and Theta are performed to derive certain parameters namely Relative Wavelet Packet Energy (RWPE), Root Mean Square Deviation (RMSD). These parameters are found to be useful in the detection of drowsiness from the EEG signal.

Relative Wavelet Packet Energy is defined as the energy within the wavelet packets derived from the discrete wavelet transform. The Energy within the wavelet packets is estimated over a period with a window of length 128K samples with 50 percent overlap as given in equation (1).

$$p(n) = \sum_{n=1}^{N} \sum_{k=1}^{M} X^2(n+k-1) \quad (1)$$

X (k) is the Segmented EEG data by DWT
p(n) is the RWPE of the signal
M is the length of the signal

The length of the window (128K) is chosen such that the Energy within the signal for every 2 minutes is computed with a moving window. The for all rhythms of EEG as the number of samples are different in each of them due to down sampling as shown in Fig.2.5.

The Root Mean Square Deviation (RMSD) is an aggregation of the magnitudes of the errors in predictions for several times into a single measure of predictive power (Hyndman et al., 2006). RMSD [r (n)] was estimated on the Alpha (α), Beta (β) and Theta by using equation (2).

$$r(n) = \sqrt[2]{\frac{\sum_{n=1}^{N}(x_n - \bar{x})^2}{n}} \quad (2)$$

Where, x_n denotes the EEG signal
$\bar{x}$ denotes the mean of the signal N
denotes the length of the window
n denotes the index of the samples

A window of length 128k (k = 1024) with 50% overlap was chosen to find the predictive power (PRMSD) within

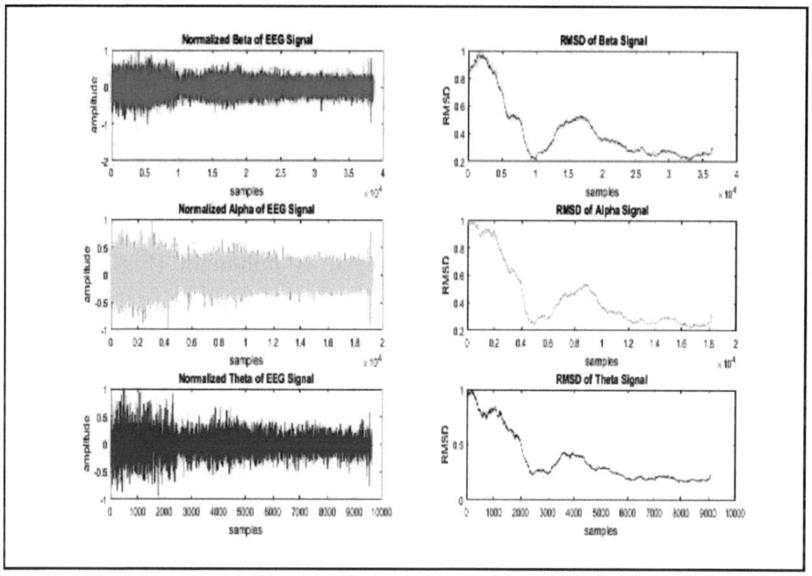

Fig.2.6. Decomposed EEG signals and their Power within the RMSD (PRMSD)

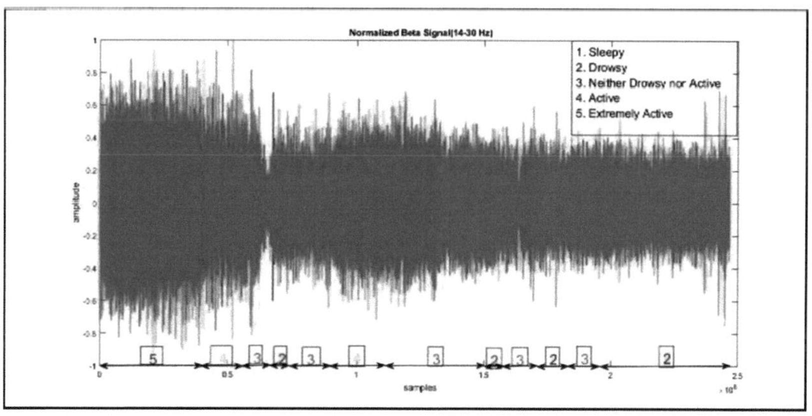

Fig.2.7. Drowsiness ratings given by the neuro-physicians based on subjective assessment

RMSD using the Eqn. 3. The predicted power (PRMSD) was normalised, and a drowsiness index was proposed in Time-Domain for Beta, Alpha and theta rhythms (Fig. 2.6).

$$q\,(n) = \sum_{n=1}^{N}\sum_{k=1}^{M} r^2(n+k-1) \qquad (3)$$

2.3. Subjective assessment

The EEG data recorded is evaluated by three neurophysicians who are real sleep experts. The experts gave the ratings, within the scale 1-5. A score of "1" as Sleepy and the score of "5" as Extremely Active. These clinical examinations correlated with the video recordings of the drivers to categorise the data into various stages of drowsiness. The Fig.2.7 shows the ratings given by the sleep experts along with normalised beta signal waveform. As can be seen from the Figure both of them correlate well with each other..Apart from this subjective assessment, both the Simulator's Driver Rating and Self-feedback by the subjects were also considered to validate the accuracy of the obtained parameters. Fig.2.7 shows the scoring given by the experts for the EEG data recorded from a subject clearly indicates the drowsiness levels ranges from 1-5 on the time axis. These ratings are used to compare and alidate the results obtained by using RWPE & PRMSD.

3. Results and discussions:

The EEG Rhythm analysis (Beta, Alpha, and Theta rhythms of temporal, parietal, occipital, and temporal lobes) is carried out through the features such as RWPE, and PRMSD Shown in Fig.2.5 & 2.6 respectively. The RWPE, PRMSD profiles of a subject clearly shows the close correlation with the subjective assessment shown in Fig.2.7.The RWPE, PRMSD are normalised, and a threshold of 0.5 is found to be the discriminating factor between the Active and Drowsy stages. The RWPE, PRMSD values of 15 subjects compared to the subjective assessment ratings to quantify the drowsiness levels.

Table 3.1 shows the statistical analysis of PRMSD, RWPE values of the 15 subjects who are under simulated driving condition. The power within the Beta, Alpha, and Theta decreases significantly in the drowsy condition across all the subjects. The cortico-neural activity of the brain decreases when a subject is performing a monotonous activity for longer periods of time. The drowsiness index proposed in concern with RWPE and PRMSD is clearly distinct in both active and drowsy cases of all the subjects, and it is statistically significant ($p < 0.05$) as the deviation from the mean is minimum (Fig.3.1). This same applies for the

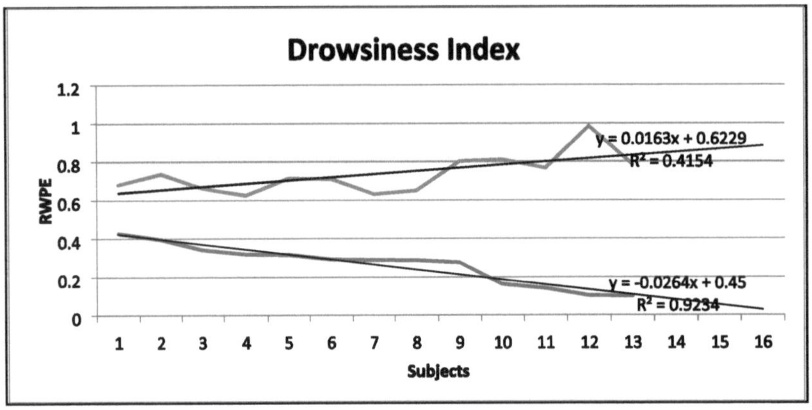

Fig.3.1. Drowsiness Index from the RWPE

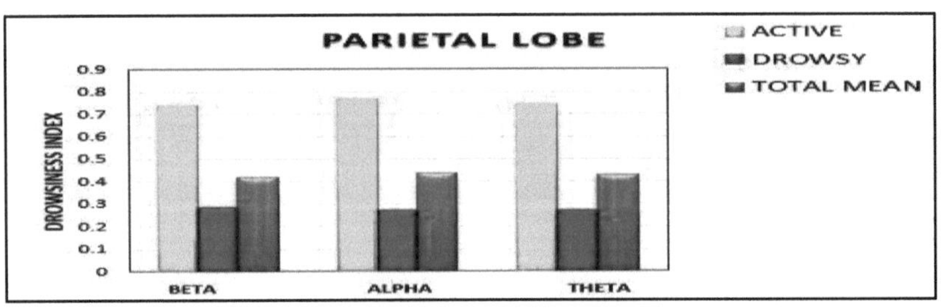

Fig. 3.2. Drowsiness index of Alpha, Beta, and Theta of Parietal lobe.

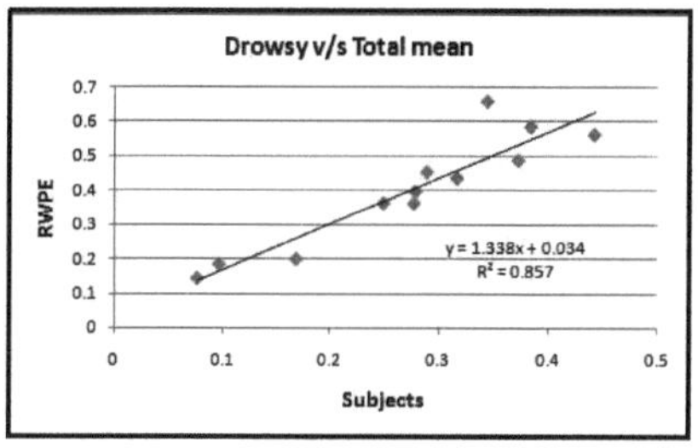

Fig.3.3. RWPE of Drowsy Mean Vs Total Mean

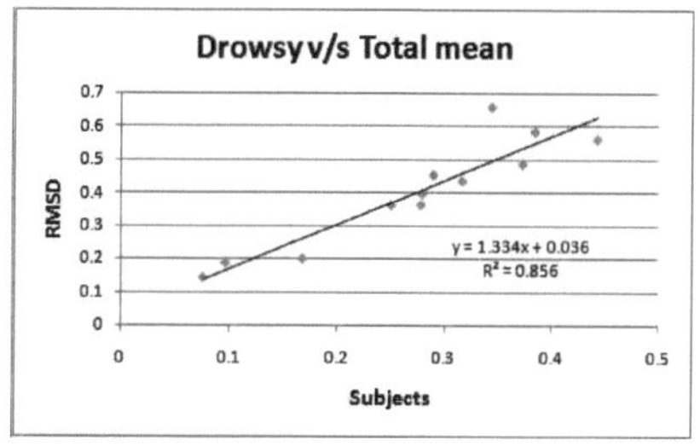

Fig.3.4. PRMSD of Drowsy Mean Vs Total Mean

PRMSD, and there exists a clear distinction between the active and drowsiness levels as can be seen in Fig.3.1.

Table.3.1. Drowsiness index derived from PRMSD and RWPE

DROWSINESS INDEX		PRMSD	RWPE
ALPHA	*ACTIVE*	0.741±0.020	0.742±0.020
	DROWSY	0.265±0.001	0.265±0.001
BETA	*ACTIVE*	0.736±0.054	0.737±0.054
	DROWSY	0.265±0.002	0.265±0.003
THETA	*ACTIVE*	0.733±0.038	0.734±0.038
	DROWSY	0.249±0.002	0.249±0.002

Regression analysis of RWPE and PRMSD shows that there exists a significant correlation ($p< 0.05$) in the detection of drowsiness with Confidence Interval of 14. This study also reports a significant correlation ($p< 0.05$) between RWPE of Total mean with Active mean R^2 of 0.857.A regression line between the PRMSD of Total mean and the drowsy mean means, it shows the close correlation of R^2 of 0.856. The drowsiness index derived from the RWPE and PRMSD is found to be less than 0.5 for drowsy state and greater than 0.5 for the active condition. The Drowsiness index constitutes significant correlation for rounded four lobes namely temporal, parietal, occipital, and frontal. The power of the signal is reduced in Alpha, Beta, and Theta rhythms during the drowsy condition are clearly indicated in Fig.3.3.

Figure3.3 (a, b) shows the correlation between the total mean of drowsiness index derived from RWPE and PRMSD versus Drowsy mean. The study shows a positive correlation of R^2 of 0.83-0.96 between the total mean and Drowsy mean, which is significant ($\rho < 0.05$) with a confidence interval of 11.

4. Conclusion:

Drowsiness detection through EEG based statistical parameters is executed on a virtual driving environment. The PRMSD of subjects shows there is a significant change in the attention levels of all the subjects. The subjects conditioned for long term monotonous driving tasks results in the decrease of the cortico-neural activity. Wavelet packet decomposition (7- levels) of Beta, Alpha, and Theta has been done and two parameters PRMSD & RWPE for total mean and drowsy mean are found to be significant with a positive coefficient of correlation 0.856 and 0.857. The total driving profiles of the subjects are classified into Active and Drowsy and are validated by using subjective assessment. Drowsiness index has been derived from PRMSD and RWPE of Beta, Alpha and Theta. It is observed that the drowsiness index from PRMSD, RWPE in active and drowsy stages of Alpha and Beta are statistically significant. A relation is proposed between the Total mean of PRMSD, RWPE and Drowsy mean of PRMSD, RWPE for Alpha and Beta. There is a strong correlation between 0.83-0.92 the total mean and drowsy mean of PRMSD and PSG. Among the two methods RWPE is easy to implement in real time compared to the PRMSD as it requires many computations. Though the study proposed a novel method for detection of drowsiness the validation of it in real time scenario yet to be explored. Wireless sensors with Real-Time analysis on DSP processor would be a potential tool for the commercial applications.

Acknowledgements:

The authors wish to thank the neuro physicians of Apollo hospitals, Hyderabad, India for their valuable contributions in subjective assessment.

References:

1. CONNOR. J., R. NORTON, S. AMERATUNGA, E. ROBINSON, I. CIVIL, R. DUNN, J. BAILEY, R. JACKSON, "Driver sleepiness and risk of serious injury to car occupants: population based case control study" Br. Med. J., 324 (7346) (2002), p. 1125.
2. Roads and Traffic Authority, NSW,2011, "Heavy truck crash data analysis: single vehicle crashes, NSW Centre for Road Safety, RFAC Meeting, September" http://www.rta.nsw.gov.au/roadsafety/downloads/accident statistics dl4.html (2011).
3. BELYAVIN, A., WRIGHT, N.A.,1987 "Changes in electrical activity of the brain with vigilance, Electroencephalography and Clinical Neurophysiology, 1987,66,137-144.
4. *KECKLUND, G., AKERSTED, T.,1993,*"Sleepiness in long distance truck driving: An ambulatory EEG study of night Driving", *Ergonomics, Vol.36, pp, 1007-1017.*
5. CHIN-TENG LIN., RUEI-CHENG WU., TZYY-PING JUNG., SHENG-FU LIANG., TENG-YI HUANG.,2005. *"Estimating Driving Performance Based on EEG Spectrum Analysis".,* EURASIP Journal on Applied Signal Processing 2005:19, 3165-3174.
6. JAGANNATH M, BALASUBRAMANIAN V, 2014. Assessment of early onset of driver fatigue using multimodal fatigue measures in a static simulator. Applied Ergon. 2014 Jul;45(4):1140-7. DOI: 10.1016/j.apergo.2014.02.001. pub: 2014 Feb 26.
7. T raffic Index for country,2016 (http://www.numbeo.com/traffic/rankings by country.jsp)
8. VANDNA SAINI AND REKHA SAINI, "Driver Drowsiness Detection System and Techniques" IJCSIT, Vol. 5 (3), 2014, 4245-4249. Road Safety Information, Rospa, "Driver fatigue and Road accidents." (www.rospa.com, 2011)
9. The National Highway Traffic Safety Administration (http://www.nhtsa. gov/).
10. Real Time Traffic Accident Statistics (http://www.icebike.org/real-time-traffic-accident-statistics/)
11. J.A. HORNE, L.A. REYNER." Driver sleepiness ", J. Sleep Res., 4 (S2) (1995), pp. 23-29
12. National Transportation Safety Board, 2010. Truck- Tractor semitrailer Rear-End Collision into passenger vehicles on Interstate 44, Near Miami, Oklahoma, June 26, 2009, Highway Accident Report, NTSB/HAR10/02, Washington, DC.
13. GRENECHE, J., KRIEGER, J., ERHARDT, C., BONNEFOND, A., ESCHENLAUER, A., MUZET, A. and TASSI, P., 2008. EEG spectral power and sleepiness during 24h of sustained wakefulness in patients with obstructive sleep apnea syndrome. *Clinical Neurophysiology,* 119(2), pp. 418-428.
14. SAHAYADHAS, A., SUNDARAJ K. and MURUGAPPAN, M., 2013. Drowsiness detection during different times of day using multiple features. *Australasian Physical& Engineering Sciences in Medicine,* pp. 18.
15. JI, Q. and YANG, X., 2002. Real-time eye, gaze, and face pose tracking for monitoring driver vigilance. *RealTime Imaging,* 8(5), pp. 357-377.
16. PAPADELIS, C., CHEN, Z., KOURTIDOU-PAPADELI, C., BAMIDIS, P.D., CHOUVARDA, I., BEKIARIS, E. and MAGLAVERAS, N., 2007. Monitoring sleepiness with onboard electrophysiological recordings for preventing sleep-deprived traffic accidents. *Clinical Neurophysiology,* 118(9), pp. 19061922.
17. LIU, N., CHIANG, C. and HSU, H., 2013. Improving driver alertness through music selection using a mobile EEG to detect brainwaves. *Sensors,* 13(7), pp. 8199-8221.
18. HOMAN, R.W., HERMAN, J. and PURDY, P., 1987. Cerebral location of international 10-20 system electrode placement. Electroencephalography and Clinical Neurophysiology, 66(4), pp. 376-382.
19. BAREA, R., BOQUETE, L., MAZO, M. and LOPEZ, E., 2002. System for assisted mobility using eye movements based on electrooculography. *Neural Systems and Rehabilitation Engineering, IEEE Transactions on,* 10(4), pp. 209-218.
20. YOUNG, L.R. and SHEENA, D., 1975. Eye-movement measurement techniques. *American Psychologist,* 30(3), pp. 315.
21. RODRIGUEZ-IBANEZ, N., GARCIA-GONZALEZ, M., FERNANDEZ-CHIMENO, M. and RAMOS- CASTRO, J., 2011. Drowsiness detection by thoracic effort signal analysis in real driving environments, Engineering in Medicine and Biology Society, EMBC,2011 Annual International Conference of the IEEE *2011, IEEE, pp. 6055-6058.*

State space modeling of bioimpedance for haematocrit measurement

S.A. Akulov[1], A.A. Fedotov[1], A.S. Akulova[2], M.A. Reshetnikova[3]

[1] Samara State Aerospace University, Department of Laser Systems and Biomedical Engineering, associate professor, Samara, Russia
[2] Samara State Aerospace University, Department of Laser Systems and Biomedical Engineering, Ph.D. student, Samara, Russia
[3] Samara State Aerospace University, Department of Laser Systems and Biomedical Engineering, graduate student, Samara, Russia

Abstract— **Haematocrit are recorded during artificial dialysis resulting in remarkable changes in the blood conductivity that can reach up to 20%. Traditional methods for HCT-measurement like centrifugation or photometry are well known and precise. The drawback of the first method is the need for extracting a blood sample out of the closed loop system, which takes time and increases the cost. The latter method requires an optical window to access the blood and therefore it does require an additional precision fabricated disposable component. In order to eliminate the disadvantages of the mentioned HCT- measurement methods a new approach for blood bioimpedance modeling is presented.**

The principles of functional identification of electrical impedance of biological tissues are considered. To determine the frequency characteristic of the electrical impedance the method of transient functions is proposed. Transient function of electrical impedance recorded as a response on unit step-current. The frequency characteristic is defined as the Carson—Heaviside transform of the electrical impedance transient function. Functional identification realized by Levy method.

For modeling the component analysis of a biological tissue the electrical impedance frequency characteristic is considered in the state space of parallel RC circuits, where the state variable has the meaning of electrical charge. The modeling of electrical impedance in blood samples with different haematocrit level is carried out. The dependences of the equivalent circuit parameters upon the haematocrit level are obtained. The obtained dependences of the impedance modulus characteristic frequency and the equivalent scheme parameters upon the haematocrit level may be used for the haematocrit level determination in a blood sample.

Keywords— **bioimpedance, electrical equivalent circuit, frequency characteristic, state-space model, transient response.**

I. Introduction

The study of biological tissue properties that determine their ability to conduct electrical current is an important field of biophysical research, the results of which are widely used in the impedance spectrometry of biological objects and non-invasive methods of medical diagnostics, such as reography, impedance cardiography and tomography [1]. An urgent line of development in the studies of bioelectrical impedance is the creation of methods for express-evaluation of the structural composition of biological tissues, e.g., the fraction of fat tissues or free fluid in the organism [2], as well as the assessment of haematocrit level in blood samples [3].

Bioimpedance methods of measuring tissue electrical properties are based on the determination of amplitude-frequency characteristics of the bioelectrical impedance parameters. The determination of the frequency characteristic of bioelectrical impedance is often implemented by measuring the components of the complex electrical impedance of the tissue at different frequencies of the alternating electrical current. For this aim a sinusoidal current from a tuneable-frequency generator is supplied to the fragment of biological tissue under study, and the amplitude and phase measurements of the voltage drop at this subcircuit at different frequencies. Using the obtained data the determination of the bioelectrical impedance in frequency domain is performed. However, in this case it is difficult to acquire complete information about the frequency characteristics of the bioelectrical impedance, first of all, because of the limited number of frequencies at which the measurements are carried out, as well as because of the requirements of using small amplitudes of acting current.

II. Theory

To obtain the frequency characteristic of the bioelectrical impedance it is reasonable to use the method of functional identification of systems [4]. The functional identification (Figure 1) is implemented by testing the object under study using the impacts that allow the determination of its transfer function, relating the object response with the exerted action. In the identification of systems the stepwise, pulsed, or noise action is used as a testing one [5]. The analysis of the transfer function allows the determination of frequency characteristic of the object with respect to the class of used testing actions.

Assume that in the functional identification of bioelectrical impedance the electrical current is used as the input

J. Goh et al. (eds.), *The 16th International Conference on Biomedical Engineering*,
IFMBE Proceedings 61,
DOI: 10.1007/978-981-10-4220-1_18

testing action, and the recorded response is the voltage that appears at the considered subcircuit. Then the transfer function describes the time variation of the bioelectrical impedance, and its processing yields the impedance frequency characteristic.

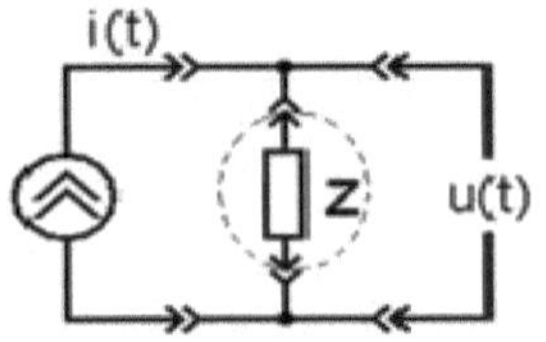

Fig. 1. Functional identification of bioelectrical impedance

The following conditions are considered as the basic assumptions in the modeling of bioelectrical impedance:

-the configurations and the parameters of the electrical impedance of the studied biological object do not vary in time;

-the characteristics of the bioelectrical impedance of the studied biological objects possess the property of linearity, i.e., the superposition principle is valid for them;

-the biological object exhibits the property of dissipativity under the action of the electrical current.

Let us use the Heaviside method [6] to find the frequency characteristic of the bioelectrical impedance. We make use of the decomposition of the input action into components by beams of the unit Heaviside function having the form:

$$\gamma(t)=\begin{cases}1, t\geq 0\\ 0, t<0\end{cases} \tag{1}$$

When the step function of the form $\gamma(t)$ is applied to the input of the object, the output reaction of the object is presented by the transfer function $g(t)$ characterizing the response temporal behavior. Let us present the arbitrary current described by a continuous function and used as the input action in the form of the expansion:

$$i(t)=i(0)\cdot\gamma(t)+\Delta i_1\cdot\gamma(t-\Delta t)+\ldots+\Delta i_n\cdot\gamma(t-n\cdot\Delta t)$$

where $\gamma(t)$ is the Heaviside function.

The voltage on the studied object will be determined as a sequence of reactions of the object to the sequence of Heaviside functions:

$$u(t)=i(0)\cdot g(t)+\Delta i_1\cdot g(t-\Delta t)+\ldots+\Delta i_n\cdot g(t-n\cdot\Delta t)$$

Making the increments to be infinitely small, we get:

$$u(t)=i(0)\cdot g(t)+\int_0^t g(t-\tau)\cdot i'(\tau)d\tau \tag{2}$$

For the operator impedance, introduced as a ratio of the operator voltage to the operator current, i.e., representing the transfer function of the system, we arrive at the expression that presents a Carson transform of the transfer function of the object under study:

$$Z(p)=p\cdot\int_0^\infty g(\tau)\cdot\exp(-p\tau)d\tau \tag{3}$$

For $p=j\omega$

$$Z(j\omega)=j\omega\cdot F\{g(t)\} \tag{4}$$

where F{$g(t)$} is the Fourier transform of the transfer function $g(t)$.

The expression (3) allows determination of the operator impedance of the studied biological object in the form of the integral Carson transform. Therefore, the frequency characteristic of the biological object electrical impedance can be determined step by step using the scheme “testing by stepwise current- recording the transfer function – the transformation (4)”.

To assess the structural composition of biological tissues the bioelectrical impedance is presented in the form of an equivalent circuit [7]. Resistive component of the impedance characterize the current propagation in the electrolyte media (blood plasma, lymph, interstitial fluid, etc.) and are related to the charge transfer in the electrical circuit. The capacitive components of the impedance are related to charge separation phenomena that occur in multicomponent media due to polarization currents. If the electrical properties of the object structure element are presented by a cell, consisting of a parallel connection of the active resistor Ri and the capacitor Ci, then the description of the relation between the cell voltage $Ui(t)$ and the current $I(t)$, passing through the considered element, is determined by the equation:

$$R_iC_i\frac{dU_i(t)}{dt}+\frac{U_i(t)}{R_i}=I(t) \tag{5}$$

The corresponding operator impedance has the form

$$Z_i(p)=\frac{R_i}{1+pR_iC_i} \tag{6}$$

The electrical properties of a multicomponent tissue are determined by a high-order differential equation, which can be reduced to a set of first-order differential equations using the state space method. In accordance with this method [8] the variables characterizing the state of the system are introduced. These variables form the state vector that describes the behavior of the system. Such approach allows the most complete description of the modeled system, and in many cases using the methods of structural identification, the representation of underlying mechanisms.

Thus, a multicomponent tissue is modeled by the equivalent circuit, including a series connection of the active resistor $R0$ and N parallel cells $RiCi$ with the operator impedance:

$$Z(p)=R_0+\sum_{i=1}^{n}\frac{R_i}{1+pR_iC_i} \quad (7)$$

Introducing the state variable

$$S_i(p)=\frac{I(p)}{p+\frac{1}{R_iC_i}} \quad (8)$$

and transforming to originals, we get the set of state equations describing the electrical properties of the model

$$\begin{cases}\dot{S}=AS+BI\\ U=CS+DI\end{cases} \quad (9)$$

where I is the input action (current) vector, U is the output response (voltage) vector, S is the state vector, A,B,C,D are matrices of coefficients.

$$A=\begin{bmatrix}1/R1C1 & \dots & \dots & 0\\ 0 & 1/R2C2 & \dots & 0\\ \dots & \dots & \dots & \dots\\ 0 & \dots & \dots & 1/RnCn\end{bmatrix}, S=\begin{bmatrix}s_1\\ \vdots\\ s_n\end{bmatrix} B=1,$$

$$C=[1/C_1\dots 1/C_n],\ D=R0$$

It is easy to understand that the state variable here has the physical meaning of the electrical charge, accumulated in the elementary cell. Modeling diagram corresponding to equations (7) is shown on Figure 2.

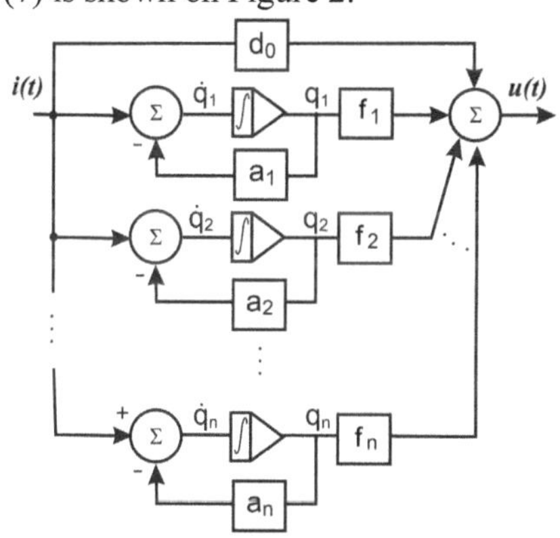

Fig. 2. State-space model of the bioimpedance

To identify the bioelectrical impedance model it is sufficient to transform the frequency characteristic obtained as a result of the experimental data processing by calculating $Z(j\omega)$, into the expression for the transfer function (e.g., by means of the Bode or Levi methods [5]). Then, in correspondence with the model, the transfer function is presented as a fractionally rational expansion, and by equating the constant coefficients the values of the model *RiCi* parameters are obtained. The interpretation of the model parameters yields the information about the structural composition of the studied object.

III. Results

This technique was used to estimate the blood haematocrit level. To begin the study, the electrodes forming a tetrapolar measuring system are immersed by the fixed depth in the measuring cell containing the blood to be analyzed. The measuring current from the appropriate generator is supplied to the electrodes and propagating through the blood gives rise to the voltage drop, recorded by an amplifier. From the amplifier output the voltage that carries information about the value of the studied blood impedance comes to the microcontroller, where the total complex resistance of the studied blood is calculated.

The resulting frequency characteristics for blood samples with different haematocrit level are shown in Figure 3.

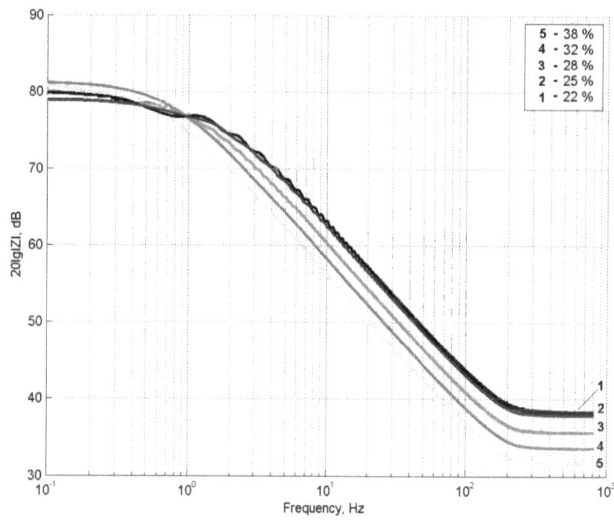

Fig. 3. Frequency characteristics of the impedance modulus for the blood samples with different haematocrit level: 1 – 22 %; 2 – 25 %; 3 – 28%; 4 – 32 %; 5 – 38 %

The dependence of blood impedance upon the frequency is complex. For "low" frequencies (<1 Hz) the impedance absolute value amounts to a few kiloohm and weakly depends upon the frequency. For "medium" frequencies (1 – 20 Hz) the impedance falls to the level of tens of ohms, and then in the region of "higher" frequencies the impedance demonstrates minor variation.

In the process of synthesizing the equivalent circuit of the bioelectrical impedance model the number of variables in the model (7) is determined by the number of parallel Ri – Ci connections. To evaluate the haematocrit level the blood is presented as a two-component system "haematocrit

– environment". With the appropriate choice of approximation for the measured frequency characteristic, the model consists of three elements (Figure 4).

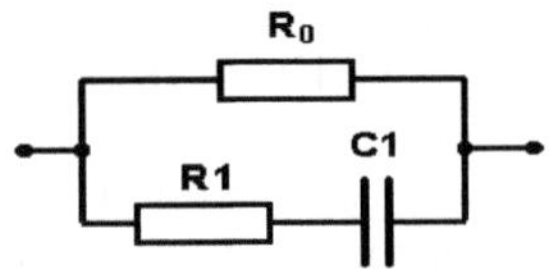

Fig. 4. Electrical equivalent circuit for the bioelectrical impedance

The parameters of this model for the obtained frequency characteristics are summarized in Table 1.

Table 1. Parameters of the bioimpedance model for the different haematocrit level

Haematocrit, %	Parameters		
	R0, Ω	*R1, kΩ*	*C, μF*
23	86	8.9	6.6
25	64	10.1	7.1
28	56	11.2	7.4
32	48	11.9	9.1
38	40	12.5	10.9

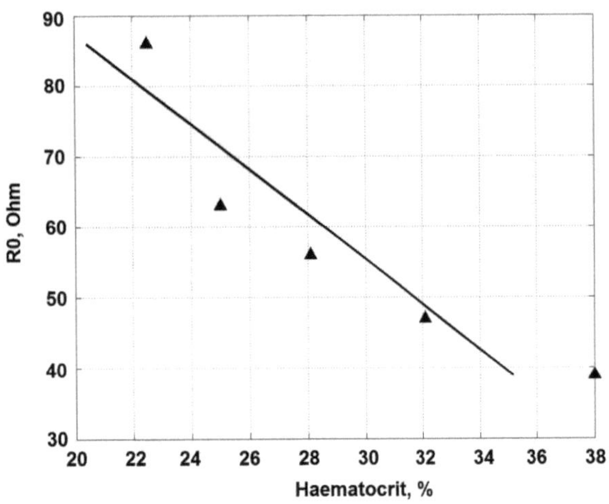

Fig. 5. Blood haematocrit level versus parameter R0

IV. Conclusion

The obtained dependence (Figure 5) of the impedance modulus characteristic frequency and the equivalent scheme parameters upon the haematocrit level may be used for the haematocrit level determination in a blood sample.

Modeling of the bioelectrical impedance of haematocrit by state space method has shown that with changing of haematocrit level there is a change of the impedance model parameters.

Conflict of Interest

The authors declare that they have no conflict of interest.

References

1. Grimnes S., Martinsen O. (2000). Bioimpedance and bioelectricity basics. Academic Press, London
2. Fors H., Gelander L. (2002). Body composition, as assessed by bio impedance spectroscopy and dual-energy X-ray absorptiometry, in a healthy pediatric population. Acta Paediatria 91: 755-760.
3. Riu P.J., Surkhy O., Bogonez P. (2005). In vivo assessment of haematocrit changes by impedance measurements. IFMBE Proc. Vol. 11(1).
4. Norman S. (2011). Systems Engineering, International Student Version, 6th Edition, John Wiley & Sons.
5. Pintelon R., Schoukens J. (2001). System Identification: A Frequency Domain Approach, IEEE Press, New York.
6. Kalakutskiy L.I., Akulov S.A. (2009). Modeling of the bioimpedance of blood by synthesis of the equivalent circuits. IFMBE Proc. Vol. 25/7: 575-577.
7. Durbin J., Koopman S. (2001). Time series analysis by state space methods. Oxford University Press, Oxford, UK.
8. Eykhoff P. (1981). Trends and progress in system identification. Pergamon Press.

Corresponding author:

Author: Fedotov Aleksandr Aleksandrovich
Institute: Samara State Aerospace University
Street: Kirova avenue, 303, #11
City: Samara
Country: Russian Federation
Email: fedoaleks@yandex.ru

A Hybrid Classifier for the Detection of Microaneurysms in Diabetic Retinal Images

Aishwarya R.[1], Vasundhara T.[2], Ramachandran K.I.[3]

[1]Department of Electronics and Communication Engineering, Amrita School of Engineering, Coimbatore, Amrita Vishwa Vidhyapeetham, Amrita University, India.
aishwaryar36@mail.com
[2]Department of Electronics and Communication Engineering, Amrita School of Engineering, Coimbatore, Amrita Vishwa Vidhyapeetham, Amrita University, India.
[3]Centre for Excellence in Computational Engineering and Networking, Amrita School of Engineering, Coimbatore, Amrita Vishwa Vidhyapeetham, Amrita University, India.

Abstract: Diabetic Retinopathy (DR) is a chronic, progressive ocular disease in which the human retina is affected due to an increasing amount of insulin in blood. The prevalence and incidence of DR is associated with people having prolonged hyperglycaemia and other symptoms linked with diabetes mellitus. DR, if not detected and treated in time poses threat to the patient's vision ultimately causing total blindness. Among the various clinical signs, microaneurysms (MAs) appear as the early and first sign of DR. The accurate and reliable detection of microaneurysms is a challenging problem owing to its tiny size and low contrast. Successful detection of microaneurysms would be more useful for a proper planning and appropriate treatment of the disease at the early stage. The work mainly envisages the improvement of the classification accuracy by employing a hybrid classifier which combines Support Vector Machine (SVM), Naïve Bayes Classifier and the decision tree. In contrast to many other classifiers the proposed classifier works efficiently, proves to be simple in terms of computational complexity and also gives good results. The performance is evaluated using publicly available retinal image database DIARETDB1.The hard decision fusion among the three classifiers carried out using the majority voting rule gives accuracy, sensitivity and specificity of 82.2916%, 82.692%, 81.818% respectively.

Keywords: Diabetic Retinopathy. Microaneurysm. Support Vector Machine

I. INTRODUCTION

Diabetes mellitus (DM) is a chronic condition characterized by abnormally high levels of sugar in blood. It occurs either when there is an insulin deficiency or when our body becomes resistant to the insulin. When DM lasts for a period of 10 years or more, it results in a disease called Diabetic Retinopathy (DR).In India more than 62 million diabetic individuals are currently diagnosed with the disease. The prevalence of diabetes is predicted to double globally from 171 million in 2000 to 366 million in 2030 with a maximum increase in India. It is predicted that by 2030 diabetes mellitus may afflict up to 79.4 million individuals in India.

DR is an important complication of DM and may cause visual disability in diabetics as it progresses.[1]. Studies show that it is major cause of blindness and sight problems over the entire population. The incidence of DR is 50% after 10 years and 90% after 30 years of acquired diabetes [1].Some of the common symptoms of DR include spots or dark strings floating in your vision , fluctuating vision, impaired color vision, dark or empty areas in vision and sudden vision loss. The severity of the disease is determined by the number and type of lesions present in fundus image [1].

DR is classified into two stages depending upon the presence of clinical features which are Non- proliferative Diabetic Retinopathy (NPDR) and Proliferative Diabetic Retinopathy (PDR). NPDR is the beginning stage in which symptoms will not be much evident. This stage contains the early signs of presence of DR such as microaneurysm and dot hemorrhages caused by break in tiny vessels called capillaries [2]. The only way to detect NPDR is by fundus photography. PDR is the second stage where the abnormal new blood vessels which are formed at the back of the eye bursts and bleeds which finally leads to blurred vision. [2],[3].

Among the various clinical indications of DR, microaneurysm being the first sign, the proper and reliable detection of MA will aid in preventing further development

J. Goh et al. (eds.), *The 16th International Conference on Biomedical Engineering*,
IFMBE Proceedings 61,
DOI: 10.1007/978-981-10-4220-1_19

of DR. But the challenge is that MAs are small in size and similar to blood vessels in colour. So accurate and reliable detection of MAs is still an open problem. In already proposed systems using medical image processing techniques, various segmentation and classification methods were discussed for the purpose. Many of which are quite complicated and time consuming.

In this work, a hybrid classifier for the detection of microaneurysms in diabetic retinal images has been presented. It is a fusion of three classifiers and the results were verified by applying hard decision rule.

Further the work is organized as follows. Part I presents the motivation and objective, II covers the literature review related to this work, III explains the methodology and IV shows the results obtained. Section V briefly provides a conclusion to this work and finally VI discusses the scope of the work and any possible enhancements of the work done.

II. PREVIOUS WORK

In [4], a hybrid method for the detection of fine microaneurysms from non-dilated diabetic retinopathy retinal images is discussed. In this work, the first level of segmentation was carried out using morphological operations and the second level of segmentation using Naive Bayes classifier. In [5], an ensemble-based framework was proposed to improve microaneurysm detection. In this work, a combination of internal components of microaneurysm detectors, namely preprocessing methods and candidate extractors was used. The algorithm gave good results with a promising AUC.In [2], a hybrid classifier was presented which used a combination of Gaussian mixture model (GMM), support vector machine (SVM) and an extension of multimodel mediod based modeling approach in an ensemble to improve the accuracy of classification. The method proposed in [6] realizes MA detection through the analysis of directional cross-section profiles centered on the local maximum pixels of the preprocessed image. Naïve Bayes classification was employed to exclude spurious candidates. In [7], a new approach to the computer aided diagnosis (CAD) of diabetic retinopathy was discussed. This approach is based on multi-scale correlation filtering (MSCF) and dynamic thresholding.

III. METHODOLOGY

The retinal images are taken from DIARETDB1 database. In the following section, the various steps like preprocessing, segmentation feature identification and extraction and classification are explained in detail.

Preprocessing

Non uniform illumination, noise, low contrast, pigmentation and differences in camera are some of the factors that degrades the quality of an image. The preprocessing helps in the attenuation of such image variations and improves the quality of the image. The green plane (Ig) of the RGB image is extracted as the vessels and MAs are best represented in green channel of the image. A median filtering operation is performed on Ig for the noise attenuation. In median filtering, the median value is one of the pixels in the neighborhood, it doesn't create a new pixel for replacement. Contrast limited adaptive histogram equalization (CLAHE) is performed on the median filtered image after which the dark region in the image gets dominated. Then, a shade correction algorithm is applied on the image in order to avoid the variation of brightness across the image.

Exudates and Vessel elimination

Exudates are bright lesions that have to be removed prior to the process since there are chances that cause false positives of MAs due to their similarity in appearance. The other element which has to be eliminated prior to the MA detection are blood vessels. This is because, both MA and vessels are similar in appearance and as per the clinical information, there are no chances for MAs to occur on vessels. Morphological closing is the method chosen for vessel removal.

Segmentation using extended minima transform

Image segmentation simplifies an image into something that is more meaningful and easier to analyze. The segmentation results in a set of segments that collectively cover the entire image. In this study, extended minima transform is applied to the vessel eliminated image for the segmentation of microaneurysms. In extended minima transform, the computation of h minima transform is followed by the computation of regional minima of the result. The h minima transform suppresses all the minima in the intensity image whose depth is less than or equal to a predefined threshold. The regional minima are connected sets of pixels with the same intensity value whose external boundary pixels all have a higher value. This segmentation algorithm helps in the enhancement of microaneurysms by suppressing all other background structures.

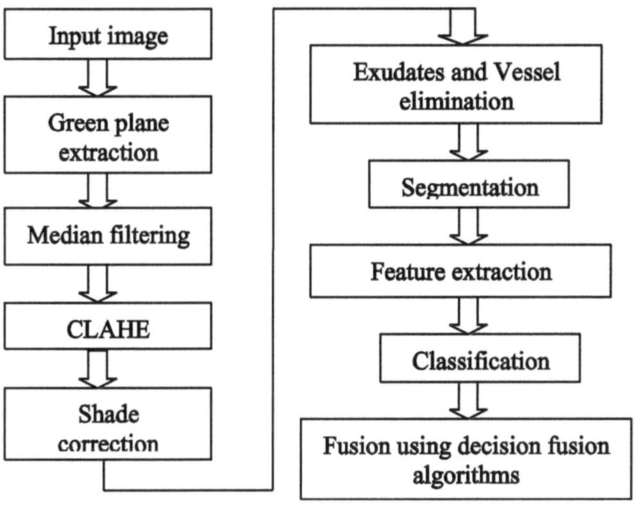

Fig. 3.2. Proposed work flow

Feature extraction

This stage helps to extract the most relevant features without much loss of information in the image. Various shape features like area, perimeter, major axis length, minor axis length, eccentricity and texture features like energy, entropy, correlation, autocorrelation, homogeneity and contrast are extracted for the classification stage.

Mathematical calculations

The features like entropy, homogeneity, energy, correlation are second order statistical features.

$$\text{Entropy} = \sum_{i=0}^{N-1}\sum_{j=0}^{N-1} Pij \log Pij \quad (1)$$

$$\text{Energy} = \sum_{i=0}^{N-1}\sum_{j=0}^{N-1} Pij^2 \quad (2)$$

$$\text{Homogeneity} = \sum_{i=0}^{N-1}\sum_{j=0}^{N-1} \frac{Pij}{1+|1-j|} \quad (3)$$

$$\text{Correlation} = \sum_{i=0}^{N-1}\sum_{j=0}^{N-1} \frac{Pij(1-\mu i)(1-\mu j)}{\sigma i \sigma j} \quad (4)$$

Pij is the occurrences of the gray level pair (i,j) where pixels are separated by some specific distance within a given neighborhood at a particular angle.

Classification

The extracted features are fed to the classifier. A hybrid classifier which is a fusion of SVM, DT and NB are used for classification in this work. The important advantage of SVM is that it can be used for both linear and non linear classification. Also, SVM works well on a wide range of classification problems, even problems in high dimensions and that are not linearly separable. The disadvantages of SVM are that the key parameters need to be set correctly to achieve the best classification results for any given problem and also, this works only for two class classification and does not work in case of multilevel classification. The other disadvantage is that SVM is computationally expensive.

Naïve Bayes classifier is a probabilistic classifier which obeys Bayes Theorem. The main advantage of this classifier is that it has a very low computational expense. But the disadvantage is that it results in loss of accuracy.

Decision tree classifier uses tree-structured algorithms for classification .The main advantages are, the classifier is robust and also simple to understand and interpret. But the main challenge is that , the criteria to decide when to stop the growing process of tree should be known.

All the three classifiers are entirely different from each other and the main idea is to bring a better result by fusing these classifiers. The result is verified using the majority voting rule.

IV. RESULTS

A total of 240 samples from 30 retinal images are used for the detection of MA in this work. Forty percentage of samples are used for testing and sixty percentage of samples are used for training in the classification stage. The algorithms developed for MA detection are written in MATLAB and are carried out on a laptop with an Intel core 2.53GHz processor and a 3GB RAM.

The results obtained from this study are evaluated quantitatively by comparing the results with ophthalmologists ground truth images quantitatively. The input images are colour retinal images of dimension (1152×1500) which are obtained from DIARETDB1_v.1.1 database as shown in the Fig. 4.1(a). As part of the preprocessing stage, green plane extraction were performed because microaneurysms and vessels have higher contrast in the green plane. Fig. 4.1(b) shows the green plane extracted image. After green plane extraction, median filtering was performed. The result of denoising is shown in Fig. 4.1(c). PSNR (Peak Signal to Noise Ratio) value were calculated for different images and it was found that PSNR value obtained after performing median filtering is high. The PSNR value obtained after and before filtering is shown in the Table 4-1. Fig. 4.1(d) shows the result after performing the Contrast Limited Adaptive Histogram Equalization (CLAHE). This helps in the enhancement of the image by highlighting certain characteristics of an image. Then, shade correction is performed in order to reduce the shading effect of the image which is caused due to the non uniform illumination. Fig. 4.1(e) represents the shade corrected image.Exudates and blood vessels have to be removed prior to the process because they can be wrongly detected as MAs due to their similarity in appearance. For eliminating exudates, morphological reconstruction is performed and is shown in Fig. 4.1(f). The blood vessels are first extracted and then are removed by the method of morphological closing which is shown in Fig. 4.1(g) and in Fig. 4.1(h).

Then, segmentation is performed using extended minima transform and the candidate microaneurysms obtained after segmentation is given in Fig. 4.2(a). After this, each of the microaneurysms from the segmented result is cropped and is fed to the next phase which is the feature extraction stage. The cropped microaneurysm is shown in Fig. 4.2(b).

The next stage is feature extraction, where features like the perimeter, area, eccentricity, major axis length, contrast, correlation, auto correlation, homogeneity, energy, minor axis length and entropy are calculated and based on the result obtained from decision tree algorithm, prominent features like perimeter, area, eccentricity, major axis length, minor axis length, homogeneity and contrast are selected and given as input to the classifiers.

Then, classification is performed using three classifiers which are Support Vector Machine, Naïve Bayes and Decision Tree. In order to improve the classification accuracy, the classifiers are fused by using a hard decision rule. Various parameters such as sensitivity, specificity, positive predictive value (PPV) and accuracy evaluate the performance of the proposed system and are shown in Table 4-1.

The results of preprocessing, exudates and vessel elimination and segmentation stages are given in the figure.

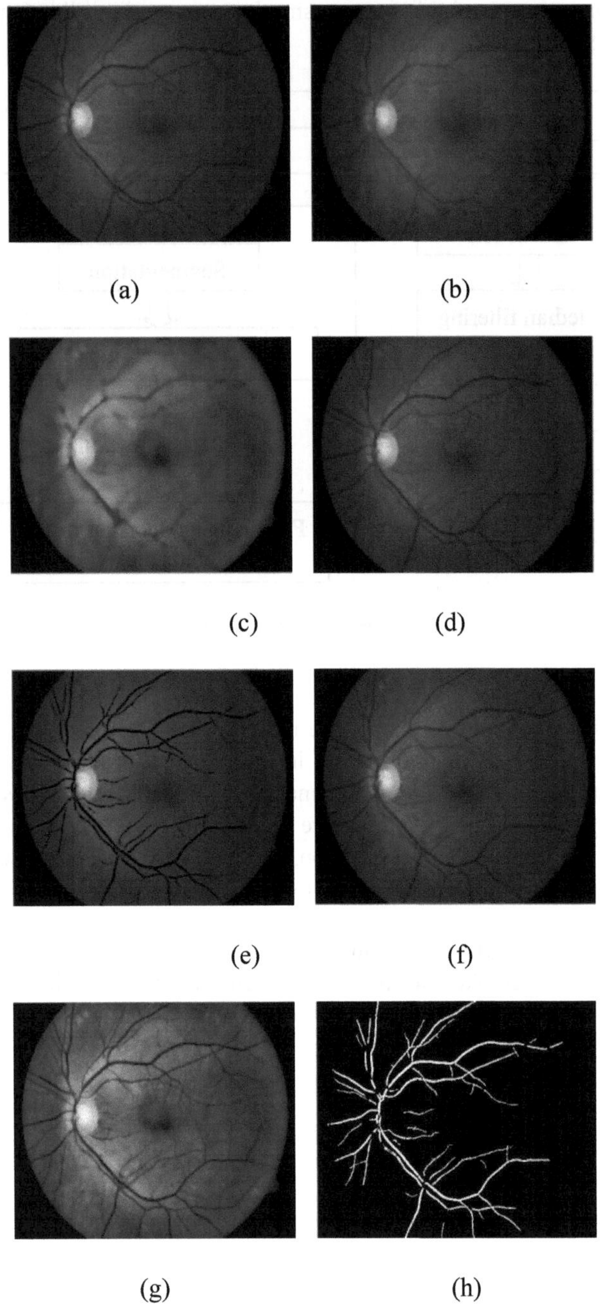

Fig.4.1. (a) original image, (b) green plane of the image, (c) median filtered image, (d) enhanced image, (e) exudate detected image, (f) extracted blood vessels, (g) vessel eliminated image, (h) segmented image

The calculation of these parameters are done by using following equations respectively:

$$Sensitivity = \frac{TP}{TP + FN} \quad (5)$$

$$Specificity = \frac{TN}{TN + FP} \quad (6)$$

$$PPV = \frac{TP}{TP + FP} \quad (7)$$

$$Accuracy = \frac{TP + TN}{TP + FP + FN + TN} \quad (8)$$

True Positive(TP) is the number of correctly detected MA pixels, False Positive(FP) is the number of non-MA pixels which are detected wrongly as MA pixels, True Negative(TN) is the number of non-MA pixels that are correctly identified as non-MA pixels and False Negative(FN) is the number of MA pixels that are not detected. The classification results obtained are as in Table.

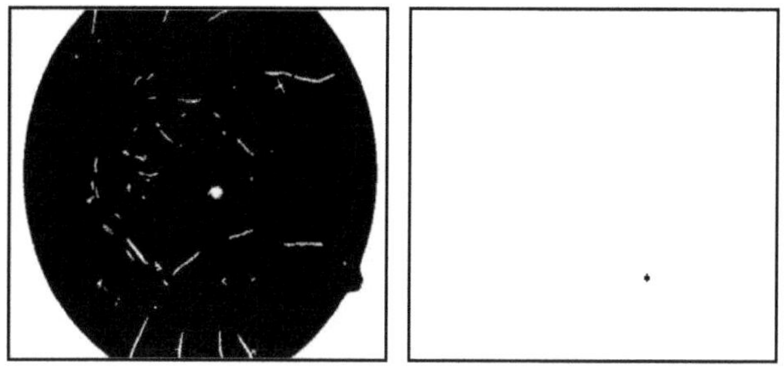

Fig.4.2. (a)Detected microaneurysms along with false positives, (b)Cropped microaneurysm (In Fig.4.2. (a) microaneurysms are detected along with false positives. So from these, true positive microaneurysms are cropped one by one and is given to the next stage which is the feature extraction stage as shown here).

Sensitivity and specificity are used to create ROC (Receiver Operating Characteristics) curve of the classifiers used in the proposed method. The ROC curve of all the three classifiers are shown in Figure.

Table 4.1. The classification results

Factors	**SVM**	**NB**	**DT**	**HC(weighted average)**	**HC(majority voting)**
TP	42	38	34	38	43
TN	28	30	35	33	36
FP	19	18	13	11	8
FN	7	10	14	14	9
Sensitivity	85.714	79.166	70.833	73.07	82.692
Specificity	59.574	62.5	72.916	75	81.818
PPV	68.85	67.857	72.34	77.55	84.31
Accuracy	72.916	70.833	71.875	73.95	82.2916

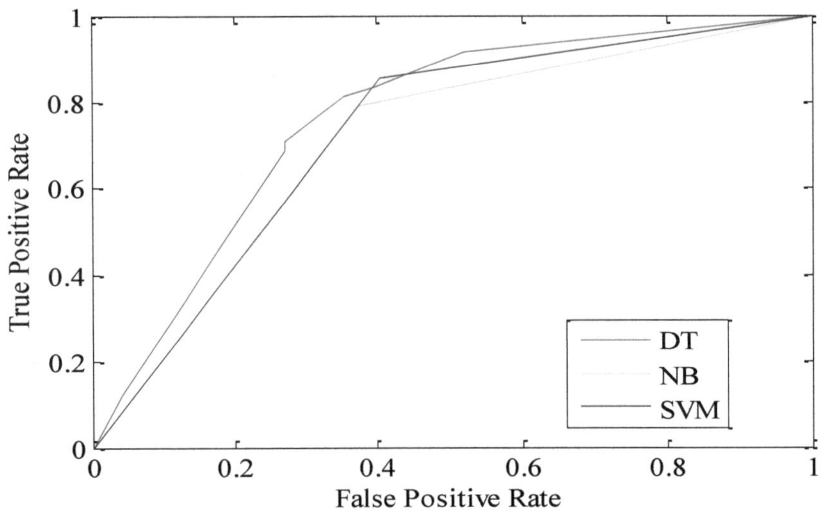

Fig 4.3. ROC curve of all classifiers

Computational time is also an important parameter for the evaluation of our system. The time taken by all the three classifiers in this work is given in the table.

Table 4.2. Time taken by the classifiers

Classifier	Computational time (s)
SVM	92.8125
NB	44.3511
DT	36.9098

V. DISCUSSIONS AND CONCLUSION

This work presented a hybrid classifier for the accurate detection of microaneurysms in diabetic retinal images. A fusion of the three classifiers namely the Support Vector Machine, Naïve Bayes and Decision Tree was employed with a view of improving the classification accuracy. Sensitivity, specificity and accuracy were used as performance metrics for evaluating the performance of the proposed system. The system produced an accuracy of 82.29% with sensitivity of 82.692% and specificity of 81.819% by the method of majority voting. The hybrid classifier employed proves to be simple as well as efficient in terms of computational complexity. In contrast to the existing state of art approaches, this proves to involve less complications and at the same time gives a comparable and good performance. The decision tree method also helped in reducing the number of features.

VI. FUTURE SCOPE OF THE WORK

The performance of the system can be improved further by selecting the most prominent features for the classification stage and also by reducing the shading effect of the retinal image. In future, the faint and small MAs can be detected by modifying the existing algorithm and also, the training time can further be reduced by reducing the feature dimension. The promising development in image resolution and computer processing power in the future allows more accurate and sensitive detection. In addition the work can be properly modified and directed towards the development of an automated system for early detection and grading of Diabetic Retinopathy (DR) using heterogeneous sources of data. The automated diagnostic system for DR can help in reducing the chances of complete blindness due to DR, along with lowering the work load on clinicians. Heterogeneous data sources including image features and personal profile data are likely to project different perspectives on the severity of the disease. These perspectives may potentially be complementary and hence a combination of these data from independent sources can aid in the holistic assessment of the disease.

REFERENCES:

[1] M. R. K. Mookiah, U. R. Acharya, C. K. Chua *et al.* "Computer-aided diagnosis of diabetic retinopathy: A review," *Computers in Biology and Medicine,* vol. 43, pp. 2136-2155, 2013.

[2] M. U. Akram, S. Khalid, S. A. Khan, "Identification and Classification of microaneurysms for early detection of diabetic retinopathy," *Pattern Recognition,* vol. 46, pp. 107-116, 2013.

[3] M. U. Akram, S. Khalid, A. Tariq *et al.* "Detection of neo vascularization in retinal images using multivariate m-Mediods based classifier" *Computerized Medical Imaging and Graphics : The Official Journal of the Computerized Medical Imaging Society,* vol. 37, pp. 346-357, 2013.

[4] A. Sopharak, B. Uyyanonvara, S. Barman, "Simple hybrid method for fine microaneurysm detection from non-dilated diabetic retinopathy retinal images," *Computerized Medical Imaging and Graphics,* vol. 37, pp. 394-402, 2013.

[5] B. Antal and A. Hajdu, "An ensemble based system for microaneurysm detection and diabetic retinopathy grading," *IEEE transactions on Biomedical Engineering,* vol. 59, pp. 1720-1726, 2012.

[6] I. Lazar and A. Hajdu, "Retinal microaneurysm detection through local rotating cross-section profile analysis," *IEEE transactions on medical imaging*, vol. 32, pp .400-407, 2013.

[7] B. Zhang, X. Wu, J. You *et al.* "Detection of microaneurysms using multiscale correlation coefficients," *Pattern Recognition*, vol. 43, pp. 2237-2248, 2010.

Approach To Extract Twin fECG For Different Cardiac Conditions During Prenatal

Rolant Gini J.[1], Ramachandran K.I.[2], Ceerthibala U.K.[1]

[1]Department of Electronics and Communication Engineering, Amrita School of Engineering Coimbatore, Amrita VishwaVidyapeetham, Amrita University, India.
jrolantgini@cb.amrita.edu, ceerthibala@gmail.com
[2] Center for Computational Engineering & Networking (CEN), Amrita School of Engineering, Coimbatore, Amrita Vishwa Vidyapeetham, Amrita University, India.
Ki_ram@cb.amrita.edu

Abstract— During multiple fetus pregnancy, degree of risk for distinguishing the information of mother and fetus health condition is high. A proper distinguishable ECG of each fetus and mother gives information about the health conditions of individuals. In case of multiple fetal conditions, the heartbeat of the fetuses will be almost at the same rate. This algorithm has been aimed to separate mECG and the fECGs of the individual fetus. First, the signal for different medical conditions like Fibrillation, Apnea, Ventricular Ectopy, Singleton and Normal has been considered. The synthetic abdECG signal for the above mentioned cases has been formulated by preprocessing and considered as the input signal. R-Peak of mECG in the abdECG signal has been located using First Order Gaussian Differentiator and Zero Crossing Detector. QRS complex has been considered around the identified R-Peak of abdECG. Identified QRS has been removed from the abdECG signal to obtain fECG with residual noise.

The QRS complexes of fECG are detected the same way as mECG QRS were detected, and is represented as binary signals. The separation of the fetal ECG is done based on the individual presence of the fetus in the signals using Inter-beat averaging and Inter-beat standard deviation of the binary signal. The algorithm has been tested for above mentioned cardiac conditions during prenatal. The algorithm has been able to achieve 99% accuracy for particular cardiac condition with overall system accuracy of 80.4%. The standard cardiac signals of different cases have been sourced from *Physionet* database to construct the abdECG.

Keywords— Twin fECG (fetal ECG), fetal Heart Rate (fHR), Cardiac Monitoring, Inter Beat Standard Deviation.

1 Introduction

The analysis of an individual is done for their wellbeing. The most important and significant health condition is known by the ECG of a person. During pregnancy, the health of Mother and Fetus is very important [9]. Extraction of ECG is one the most needed non-invasive method of analysis of the fetus for understanding its health condition, which is still an emerging area of research. If succeeded in extracting the ECG of the fetus in a non-invasive manner, it will be very useful to deliver any assistance required, preventing any kind of potential disorder or conditions, due to deterred monitoring [11]. This will improve the fetus health condition for better living and reduce the fetal or infant mortality rate. This monitoring is imperative, when the mother can have potential medical conditions like Atrial Fibrillation, Ventricular Ectopic or Apnea, etc. This can also affect the fetus health [8]. Therefore to monitor the conditions of fetal heart, the most optimum method is to detect the ECG signal of the fetus in an uncomplicated method.

The ECG signals are generally obtained by the method of electrode placement at the abdomen of the mother. These electrodes are placed in the abdomen as a set in standard positions, depending on the kind of placement required and chosen by the doctor. A typical 8-electrode system comprises of eight electrodes and a reference electrode, which allows acquirement of eight abdominal ECG channels [7]. Under multiple pregnancy condition, the recording of the abdECG signal alone, will not give an overall information about the health of the individual fetus [5]. Hence, an algorithm has to be formulated, that will not only separate the mECG signal from the abdominal recording, but will also separate the individual fECG signal, in order to monitor individual fetus.

This work aimed to develop an algorithm to identify and extract the ECG of corresponding fetus during multiple pregnancies. The maternal ECG signals are first separated from recorded abdominal signals, by the method of QRS detection and separation of the mECG. After removal of mECG, the QRS complex of the fECGs are detected, and converted to a binary signal. Separation of individual fECG signal is done by heuristic method. It is then converted to analog signal, giving the ECG of individual fetus. This

J. Goh et al. (eds.), *The 16th International Conference on Biomedical Engineering*,
IFMBE Proceedings 61,
DOI: 10.1007/978-981-10-4220-1_20

method was tested, by synthetically creating the signals. Different medical conditions of heart, such as, Atrial Fibrillation, Ventricular Ectopic and Apnea along with Singleton pregnancies, were tested by synthetic method of signal generation and separation. This paper had briefed the existing method in the literature review section, followed by the discussion of signal generation, extraction of fECG and discussion of result in the following sections. Finally the conclusion of the work had been summarized.

2 Literature Survey

The method of separation and extraction of signals, for the analysis of fetal health, is an important aspect and characteristic feature of this work. The method of analysis, as well as the method of separation, for multiple fetal pregnancies, is still an area under research. The computation for analysis has various methods.

Blind Source Separation (BSS) algorithm, is one of the separation method [12]. In this method, a series of source signals are separated from a series of mixed signals, without any or less provision of information about the source signals, by method of assimilation. There is various processes for executing the method of BSS like nonnegative matrix Factorization, imposing Structural restraints on source signals. The structural restraints model is justified by good empirical performance. This approach becomes more efficient if one does not require a whole signal, but purely the salient feature of the signal and BSS is unlikely to eliminate any desired results. This can be used with real time signals, during a multiple fetal pregnancies, under a condition, that the fECG's signals are strong enough i.e., the SNR between the noise and the fECG should be high but not possible in real time at present since fECGs are not strong signals.

Independent Component Analysis (ICA), is another computational method of separation [6]. Here a multivariate signal is separated into an additive Sub-components. It is based on the assumption that sub-components are non-Gaussian signals, that are independent of each other. ICA tries to decompose a multivariate signal into independent non-Gaussian signals. Blind ICA gives very good results when separating mixed signals, once the Statistical assumption made is correct. These assumptions also contribute to the basic Principle of the ICA analysis. It is an iterative process to extract fECG from abdECG using ICA based methods. To make ICA effective in processing time, there is advancement in the ICA and known as Fast ICA [1]. Even though it needs only one channel input for processing, the complexity in this is to decide the no of iterations requires for extracting the fECG signal. Hence, introduces difficulty in real time computation.

Like ICA, Principle Component Analysis (PCA), is another statistical method of analysis. This method gives an additional advantage in reducing the mECG interference appreciably in the analysis [15]. They are demanding in the computation process, limiting their use in separation of signals during real time analysis.

Twin fECG extraction had been done using a statistical calculation after identifying the mECG and separated by averaging and subtracting process [2]. Twin ECG had been separated where mother and fetus had normal health conditions and resulted with good accuracy and sensitivity. Even though it yields good results, the process of averaging and subtraction may not prove to be effective during prolonged or high baseline wander. It had not given with the result where there may be any cases of other cardiac conditions.

fMCG monitoring had also been tested in observing the fetal heart health in case of multiple pregnancies [5]. This observation used ICA for separation of mECG and twin fECGs and was found effective from second trimester. The fMCG device was used on the abdomen of the mother after identifying the fetus location in the womb by Ultrasonic Observation. This has been proved with good quality of extraction with a possibility of good quality retrieval of the source signal from the device. The above mentioned factor shows that it is dependent on other devices or skilled person to locate the fetuses before usage and quality of extraction strongly depends on the signal acquisition quality. The separation of the signal depends on the assumptions that there is a synchronization of QRS complexes of the beat of individual fetus, which may not be possible at every measurement.

A statistical approach proposed in this work for extraction of twin fECG for different cardiac conditions and results were discussed with acceptable accuracy levels.

3 Methodology

3.1 Generation of input ECG signals

Formation of input signals has been done by synthetic generation of signals. The abdominal signal is a result of superposition of mECG and fECG signals [10]. The algorithm assumed electrode placement as an eight-channel electrode arrangement on the mother's abdomen. The overall flow of the extraction has been summarized in Fig.1.

Two different natured signals (x1 and x2) have taken from MIT-BIH Normal Sinus Rhythm Database, available in Physio-Net [13] which was sampled at the rate of 128Hz.

One (x1) is the record of a normal individual for generating twin fetus ECG. One more record (x2) which mimics different cardiac conditions of an individual have been considered and used to generate mECG. The signals have been combined based on the nature of the mECG and fECG to generate abdECG(x) of different electrodes [2].

$$x = f_1(x1) + f_2(x1) + f_3(x2) \quad (1)$$

\where $f1$, $f2$ and $f3$ denote different functions to generate the required signal based on the signal nature and generated input signal is given by Fig. 2. The extractions of twin fECG have been studied for different cardiac conditions of the mother after generating the corresponding signals synthetically.

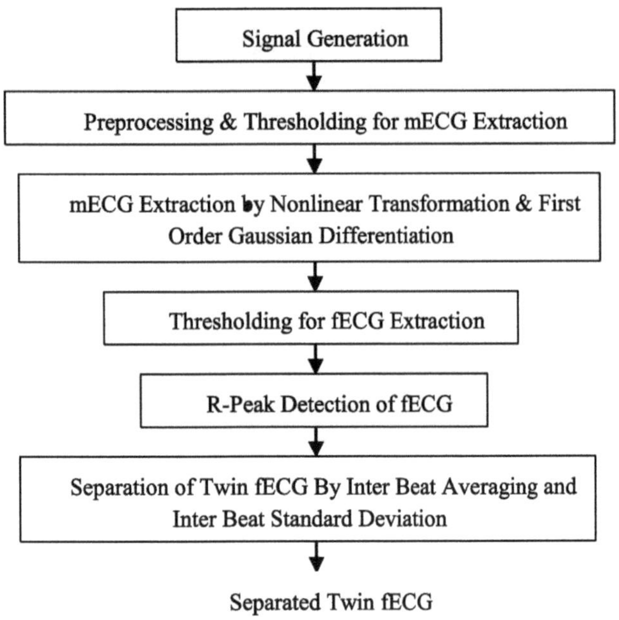

Fig. 1. Algorithm for Extraction of Twin fECG

3.2 Detection of mECG by finding R-Peaks of mother

The presence of mECG in the ECG signal will not help in the proper detection of the fECG signals. To detect mECG peaks, few major steps are followed in order to remove the mECG signal from the input signal. First is the thresholding, which helps in identifying the mECG since the amplitude of fetus will be very less when compared to the amplitude of the mother [14]. R-peak detection followed in the signal after thresholding, which is necessary for various types of ECG applications that includes identifying QRS complex, heart rate monitor (HRM), ECG wave delineator, etc., [4]. Some preprocessing has been done to adopt the signal for R-Peak detection.

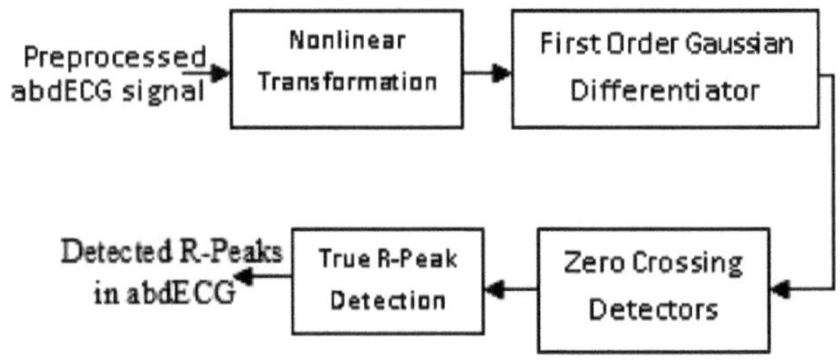

Fig. 3. mECG R-Peak Detection

3.3 Detection of R-Peaks of twin fetal ECG

Remaining signal after mECG removal is the ECG of two fetuses and the noise signal. The first step in fECG extraction is the R-peak detection of fECG. The R-peaks of fECG also undergoes the same process as R-Peak identification of mother with a change in the thresholding procedure since the nature of mECG and fECG are different [3], [14]. This resulted in identification of real fetal R-Peaks present in the signal as shown in Fig.4. and Fig.5.

These detected R-Peaks are the R-Peaks of both the fetuses present in the signal. To extract individual fetus ECG, the identified peaks have to be related only with the corresponding fetus.

3.4 Separation of Twin fetal ECG

After identification of all the fetal peaks, it had been converted into onset binary series. While converting the

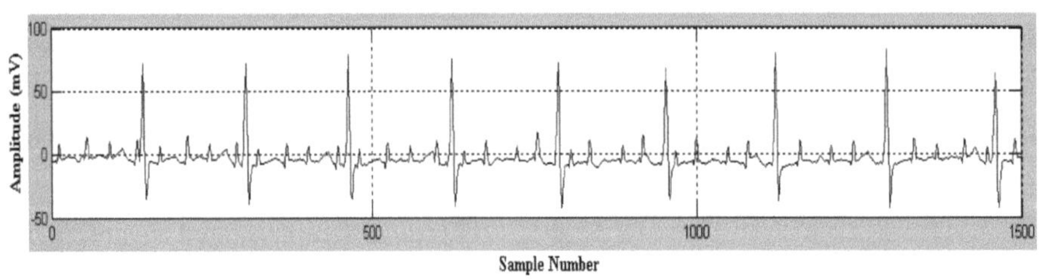

Fig. 2. Synthetically generated abdECG of a channel which contains Twin fECG

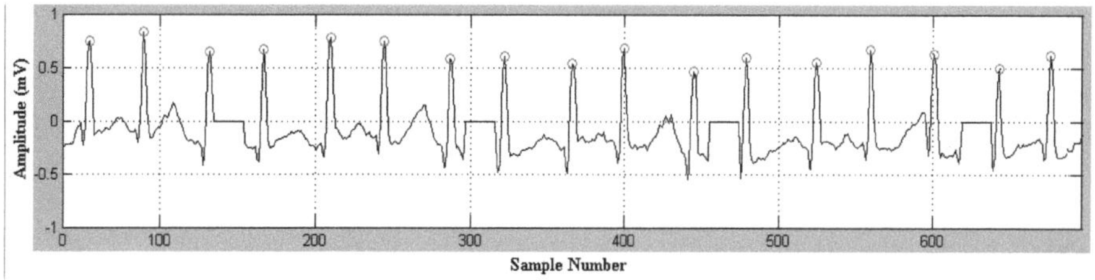

Fig. 4. mECG removed signal with fECG R-Peaks marked

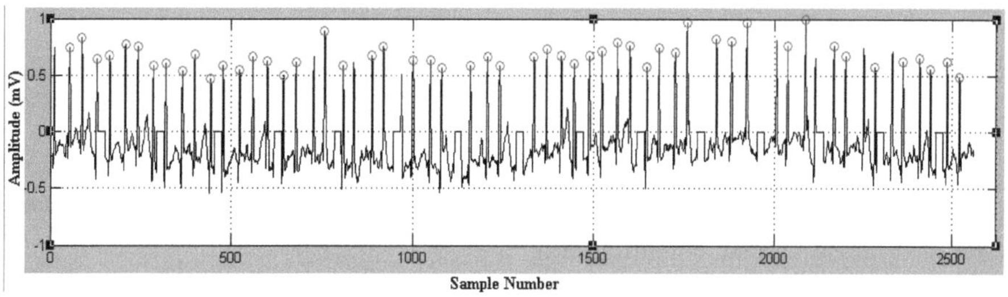

Fig. 5. Fig.5. identified entire R-Peaks of twin fECG in a channel

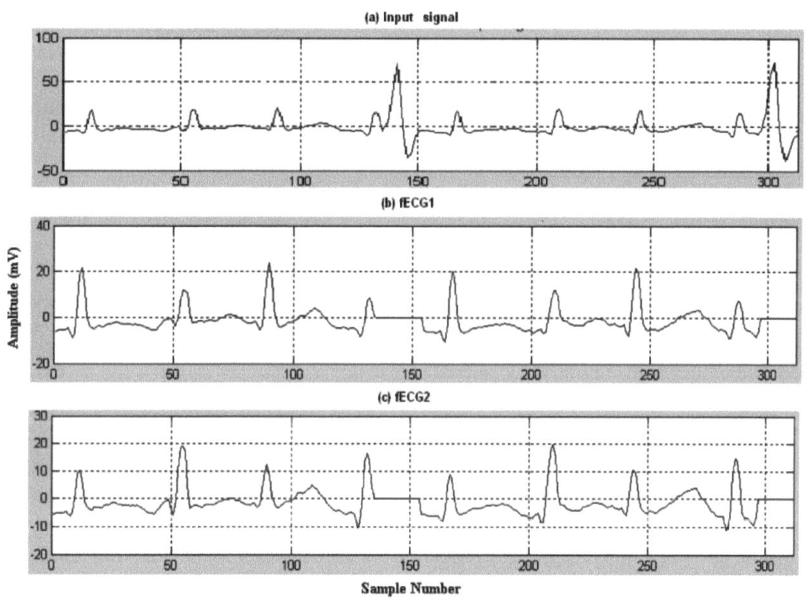

Fig. 6. (a) Input Signal (synthetic abdECG) of a channel; (b) Extracted fECG of fetus1; (c) Extracted fECG of fetus2

onset binary series, the corresponding time stamp also done by noting the sample no. All the eight channels are used for separation of twin fetal ECG. Hence after the detection of all fetal R-peaks, binary series had been obtained.

The extraction of first fetus ECG had been done based on IBAVG (Inter beat averaging) and no of R-Peaks detected in every channel. Two channels which is high in inter beat average and less in number of peak detection will be giving the ECG of one fetus [3]. Those two channels had been identified and one fetal ECG had been extracted from the same. Similarity components of the same in other six channels had been removed.

The extraction of second fetal ECG had been extracted based on IBSTD (Inter beat standard deviation) of remaining six channel signal and no of peaks detected in every channel. Two channels which show less inter beat standard deviation and number of peaks detected at least above the minimum beat required had been considered to give the

ECG of second fetus. Those channels had been identified and the fECG of second fetus had been extracted from those two channel signal. The extracted twin fetal ECG had been shown in Fig.6.

4 Results and Discussions

In this work, an algorithm to identify and extract the ECG of corresponding fetus during multiple pregnancies is presented. This algorithm was tested on synthetically generated signals. This algorithm is used to identify and separate individual fetal ECG signals in case of multiple pregnancies with effective detection of mECG, its removal and identification of fECG. The algorithm shows high dependability when it is tested for single and multiple pregnancies.

Twin ECG separation of different cardiac condition of a mother had been studied. For cases like mother having Apnoea, Atrial fibrillation, Ventricular Ectopy and Singleton pregnancy had been studied and their accuracy of separation had been tabulated in Table 1.

Table 1.

Cases	Accuracy
Apnoea	75%
Atrial fibrillation	70%
Ventricular Ectopy	92%
Singleton Pregnancy	99%

Further increasing the accuracy of the system is still possible. This algorithm is based on assumption that at least one of the leads predominantly records the heart rate of one fetus.

5 Conclusions

The algorithm was successfully tested and can be used effectively for the separation of individual fetuses in the case of medical conditions like Apnea, Ventricular Ectopy, Atrial fibrillation etc. The algorithm was tested affectively for each of the case and gave an accuracy of around 80% for each of the case .The algorithms achieved near 100% accuracy when it was tested for the singleton case. The overall accuracy of the system was found to be 80.4%.

Even if the algorithm is tested for single and twin pregnancies we believe that the same can be used for QRS detection of the fetus when more than two fetal sources are present but with a compromise in accuracy. The algorithm has certain limitations, assumptions to be tested on real time signals. The overall accuracy of this system is 80.4%. The system can be made more reliable by increasing the accuracy.

References

1. Ruben Martin-Clemente, Jose Luis camargo-Olivares, Susana HornilloMellado, Mar Elena, Isabel Roman,"Fast Technique for Non-Invasive Fetal ECG Extraction", *IEEE Transaction paper on Biomedical Engineering,* Vol.No.58, pp227-230, July 2010.
2. A.Fanelli, M.G.Signorini, T.Heldt, "Extraction of fetal heart rate from maternal surface ECG with provision for multiple pregnancies", *IEEE Conference on Engineering in Medicine and Biology,* pp-6165-6168,August 2012.
3. A. Fanelli, M.G. Signorini, M.Ferrario, P. Perego, L. Piccini, G. Andreoni, G. Magenes, "Telefetalcare: a first prototype of a wearable fetal electrocardiograph", *EMBC* 2011, pp. 6899 - 6902, 2011.
4. P. Kathirvel, M.SabarimalaiManikandan, S. R. M Prassana, K.P Soman, "An Efficient R- peak Detection Based on New Nonlinear Transformation and First-Order Gaussian Differentiator", *Cardiovascular Engineering and Technology,* Vol. 2, no. 4, pp. 408-425, December 2011.
5. S. Comani, D. Mantini, G. Alleva, E. Gabriele, M. Liberati and G.L.Romani, "Simultaneous monitoring of separate fetal magnetocardiographic signals in twin pregnancy", *Physiological Measurement,* Vol. 26, pp. 193-201, 2006.
6. M. Sridhar-Keralapura, M. Pourfathi and B. Sirkeci-Mergen, "Independent Component Analysis With Data-centric Contrast Functions For Separating Maternal And Twin Fetal ECG", *Proceedings of the World Congress on Engineering and Computer Science 2010,* pp. 0958-0966, 2010
7. M. A. Hasan et al., "Detection and Processing Techniques of fECG Signal for Fetal Monitoring", Biological Procedures Online, vol. 11, no. 1, pp. 263-295, 2009.
8. American Congress of Obstetricians and Gynecologists. Practice, Bulletin, "Intrapartum fetal heart rate monitoring: Nomenclature, interpretation, and general management principles", Tech. Rep. 109, 192 - 202, 2009.
9. R. Sameni and GD. Clifford, "A review of Fetal ECG Signal Processing: Issues and Promising Directions", The Open Pacing, Electrophysiology & Therapy Journal, vol. 3, pp. 420, 2010.
10. S. M. M. Martens, C. Rabotti, M. Mischi and R. J. Sluijter, "A robust fetal ECG detection method for abdominal recordings", *Physiological Measurements,* vol. 28, pp. 373-388, 2007.
11. C. Peters, R. Vullings, J. Bergmans, G. Oei, P. Wijn, "Heart Rate Detection in Low Amplitude Non-Invasive Fetal ECG Recordings",EMBS 2006, pp. 60926094, 2006.
12. V. Zarzoso, A. K. Nandi, E. Bacharakis, "Maternal and foetal ECG separation using blind source separation methods." *IMA J Math ApplMedBiol* vol. 14, pp. 207-225, 1997.
13. *Available online at* :http://www.physionet.org
14. Ridhu H Nair, RolantGini J, and K I Ramachandran, "A Simplified Approach to Identify the Fetal ECG from abdECG and to Measure the fHR", *IFMBE proceedings*, Vol.52, July 2015, pp 23-26.
15. Reza Sameni et al. , 'Multichannel Electrocardiogram Decomposition Using Periodic Component Analysis', *IEEE transactions on biomedical engineering,* vol. 55, no. 8, 2008.

Reception Performance Enhancement of Capacitive Micromachined Ultrasonic Transducers Via Modified Membrane Structures

Xiao Jiang[1,2], Yuanyu Yu[1,2], Jiujiang Wang[1,2], Xin Liu[1,2], U. Kin Che[1], Sio Hang Pun[1], Mang I. Vai[1,2], and Peng Un Mak[2]

[1] State Key Laboratory of Analog and Mixed-Signal VLSI, University of Macau, Macau, China
mb25551@umac.mo
[2] Department of Electrical and Computer Engineering, Faculty of Science and Technology, University of Macau, Macau, China

Abstract— This paper presents a method to enhance the reception performance of capacitive micromachined ultrasonic transducers (CMUTs) working in conventional mode and water immersion by adjusting the framework of vibrating membrane. The core conception of the enhancement is to reduce the system stiffness to improve both static membrane deflection under DC bias voltage and membrane displacement by harmonic analysis under acoustic pressure. Two modified structures of slotted membrane and corrugated membrane are presented respectively and the CMUT modeling is constructed and analyzed via finite element analysis (FEA) simulations by means of COMSOL Multiphysics. By using the modified membrane structures, FEA simulation shows that both the electromechanical coupling coefficient and receiving sensitivity are improved up to 20.9% and 50.5% for slotted membrane and 5.0% and 38.3% for corrugated membrane, respectively.

Keywords— CMUT Receiver; Sensitivity Enhancement; Slotted membrane and Corrugated membrane.

1 Introduction

Ultrasound has been widely developed and applied in many industries and medical fields such as nondestructive examination (NDE), ultrasound imaging and therapy [1]. CMUT is based on the micro-electro-mechanical system (MEMS) techniques and can solve the problem of mismatching of acoustic impedance between Lead Zirconate Titanate (PZT) and human tissue [2]. Compared to PZT, CMUT has many advantages such as low cost, wider bandwidth, easy in fabrication, better matched acoustic impedance to human body and more compatible with integrated circuit (IC) technique.

A CMUT cell can be regarded as a parallel plate capacitor which is composed of a vibrating membrane with a top electrode, a substrate also as the bottom electrode and separated by a vacuum gap. According to the diverse excited sources, a CMUT can act either as a transmitter or a receiver. Similar to output pressure of transmission CMUTs, performance of reception sensitivity is also important to the overall performance and few published papers focus on the behavior of receiving CMUTs recently [3,4]. Due to the core working principle of CMUTs receiver as a membrane forced vibration, this paper is aimed to improve the reception performance via increasing the membrane displacement.

To do this, two modified membrane structures are proposed on reception CMUTs which are slotted membrane and corrugated membrane. The transmitter CMUTs performance of slotted structure has been studied [5] and corrugated membrane is widely used in fields of condenser microphone which has a similar working principle with CMUT receiver [6]. The FEA simulation models are built via COMSOL Multiphysics (COMSOL,Inc.,Stockholm, Sweden) to analyze the CMUT receiving performance. Simulation results show the membrane deflection and output voltage across the series resistor. With these data, electrometrical coupling coefficient and receiving sensitivity can be calculated. The following section 2 shows the methodology of modified membrane structure. In section 3, FEA simulation of CMUT receiver is introduced. Simulation results and discussion are given in section 4. Finally, section 5 gives the conclusion part.

2 Methodology

For the theoretical research on the working principle of CMUT, the CMUT can be set as a forced damping vibration mechanical system [7]. CMUT works at the first resonance frequency and the resonance amplitude of membrane affects the performance of a CMUT. According to mechanical methodology, system stiffness presents the extent to which it resists deformation in response to an applied force [8]. For an ideal circular plate system with fixed rim, the flexural stiffness D is shown as (1) and the forced deflection w is expressed in (2) as a function of radial position [9].

$$D = \frac{Et_m^3}{12(1-\nu^2)} \tag{1}$$

J. Goh et al. (eds.), *The 16th International Conference on Biomedical Engineering*,
IFMBE Proceedings 61,
DOI: 10.1007/978-981-10-4220-1_21

$$w(r) = \frac{Pa^4}{64D}(1 - \frac{r^2}{a^2})^2 \quad (2)$$

Where r is the radial position, a is the plate radius, P is the external pressure, E is the Youngs modulus, t_m is the plate thickness and v is the Poissons ratio. The equations show that for a given force, a smaller stiffness will lead to a larger deflection.

For an ideal circular CMUT cell, the membrane deflection can be described in equation (2) which is related to the system stiffness shown in (1). In the slotted membrane design, some ring-shaped trenches are located at the surface of the CMUT membrane as depicted in the Fig. 1. When the trenches are constructed, the overall equivalent thickness of membrane decreases. This change will lead to the reduction of system stiffness and increasing of membrane displacement. Fig. 1 (a) shows the structure of the 4 trenches slotted membrane.

The other type of modified membrane structure is corrugated membrane which is widely applied on condenser microphone fields which have a similar working principle with CMUT receiver. In this paper, corrugated membrane is implemented by modifying the slotted structure with adding rings on the non-slotted regions between the trenches. Hence it is a rectangular-corrugated shape and 1 ring corrugated structure is related to 2 trenches, 2 rings corrugated membrane is related to 3 trenches and so on. Fig. 1 (b) shows the framework of 3 rings corrugated membrane. Similar to the slotted membrane, with the corrugated structure, membrane deflection will also increase [10].

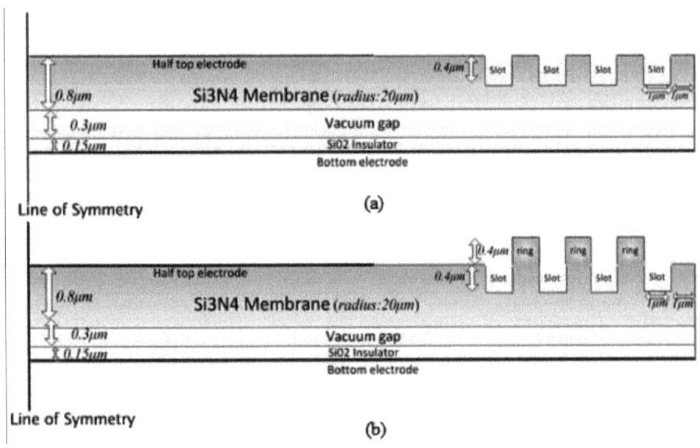

Fig. 1. Cross-sectional view and framework of CMUT receiver for (a) slotted membrane and (b) corrugated membrane.

For a receiver CMUT, membrane is biased by a DC voltage and vibrated by the effect of ultrasound pressure. When the DC bias voltage increases, the deflection of CMUT membrane also increases until the electrostatic force caused by DC voltage is larger than the restoring force of membrane, and then the membrane will collapse to the bottom plate of CMUT. This voltage is called collapse voltage [7] or pull-in voltage and a CMUT works under a voltage larger than the pull-in voltage is called collapse mode. On the other hand, a CMUT works lower than the collapse voltage is called conventional mode. This paper focuses on the property of receiver CMUTs under conventional mode due to the safety and energy cost consideration. Meanwhile, for a receiver CMUT, the collapse voltage is a very important index since the value of DC bias voltage will directly affect the reception performance of CMUTs [11]. Generally, for receiver CMUTs, DC voltage is set close to the collapse voltage and 90% of collapse voltage is used in this work. The numerical expression of collapse voltage is given by following equation (3), where k is spring constant, d is gap distance, ε is material permittivity and S is electrode area. Since the spring constant is influenced by system stiffness and less stiffness will leads to smaller k, so the collapse voltages of modified membrane structures will decreases. What is more, the membrane natural frequency is given by equation (4) where ρ is material density and other parameters are the same as that in former equations (1) (2). Due to the spring softening effect, the CMUT center frequency is smaller than the membrane nature frequency. From the equation (4), it is derived that with the modified membrane structure, the center frequency will downward shift due to the decrease of effective membrane thickness.

$$V_{collapse} = \sqrt{\frac{8kd^3}{27\varepsilon S}} \quad (3)$$

$$f = 0.47\frac{t_m}{a^2}\sqrt{\frac{E}{\rho(1-v^2)}} \quad (4)$$

The electromechanical coupling coefficient K_T^2. is defined as the ratio of delivered energy to the load over the total energy stored in the device. For the parallel plate structure approximation, it can be expressed by the following equation (5) where x is the effective displacement of top plate [12]. And t_m, ε_m are the thickness and material relative permittivity of membrane and t_i, ε_i are the thickness and material relative permittivity of insulator layer, respectively. The d_{eff} is effective gap distance of a CMUT expressed in the equation (6).

$$K_T^2 = \frac{2x}{d_{eff} - x} \quad (5)$$

$$d_{eff} = d + \frac{t_i}{\varepsilon_i} + \frac{t_m}{\varepsilon_m} \quad (6)$$

3 FEM Simulation

This part elucidates the FEM modeling process of reception CMUTs. FEM is a common tool for CMUTs simulation and the COMSOL Multiphysics presents a good performance in multiple physics modeling. According to the basic framework shown in Fig. 1 and working methodology of reception CMUT, 2D axisymmetric FEM simulation models are completed via COMSOL Multiphysics for a conventional CMUT receiver and modified membrane CMUTs. Dimensions and material parameters are shown in Table 1 and Table 2. The reception CMUT cell consists of three physics domains: *electromechanism (emi), acoustic pressure (acpr)* and *electrical circuit (cir).* The *emi* module is used to construct the CMUT cell with silicon nitride (Si_3N_4) membrane, vacuum gap and silicon dioxide (SiO_2) insulator which is set between the vacuum gap and bottom electrode to prevent electrical short circuit with the outer medium part. Half top electrode [13] is applied as the boundary condition of membrane surface and terminal type is selected as circuit to coupling with the cir physics. Moreover, 1 atm additional pressure is applied on the surface of membrane to describe the atmospheric pressure affected by the vacuum gap. The acpr module contains the outer hemispherical condition with 1000 μm radius to describe the working environment of water immersion. The boundary conditions of *acoustic load per unit area* and *normal acceleration* are set to couple the multi-physics and describe the influence of water immersion, respectively. CMUT receiver can transfer energy from mechanical energy to electrical one. Hence with the effect of acoustic pressure, electrical voltage signal is generated through the connected resistor. The *cir* physics presents the additional circuit connected to the CMUT receiver and the output voltage is calculated through the series 500 kΩ resistor.

Table 1. CMUT modeling material parameters in FEM simulation

Parameter	Value
Youngs modulus	320 GPa
Poissons ratio	0.263
Density	3270 kg/m^3
Relative permittivity(Si_3N_4)	5.7
Relative permittivity(SiO_2)	4.2

Table 2. CMUT modeling dimensions in FEM simulation

Geometry	Value
Membrane radius	20 μm
Membrane thickness	0.8 μm
Vacuum gap	0.3 μm
SiO_2 insulator thickness	0.15 μm
Center frequency (air)	9.64 MHz

Two study steps of *Stationary* and *Frequency-Domain, Perturbation* are used in COMSOL FEA simulation. For the *Stationary* study, the CMUT is deflected by a DC bias voltage and static membrane deflection is used to calculate the electro-mechanical coupling coefficient. As mentioned before, with the modified membrane structure, the collapse voltage will decrease and 90% of collapse voltage is used, so determining the value of changing collapse voltage is necessary above all when applying modified membrane structures. In the *Frequency-Domain, Perturbation* study, a 10 Pa harmonic external pressure is applied on the surface of membrane as a boundary load from 1.5 MHz to 4 MHz with a step of 10 KHz. The output voltage cross the series resistor can be measured via COMSOL and hence receiving sensitivity can be calculated.

After the FEA simulation of a conventional CMUT receiver is constructed, the next step is to replace the vibration membrane to modified structures. First of all, position consideration is given for applying one trench under different radial positions. Then up to 4 trenches for slotted membrane and at most 3 rings for corrugated membrane are presented. The details of simulation result will be shown in the Section 4.

4 Results and Discussions

First of all, position consideration is given. Then slotted membrane and corrugated membrane structures are constructed to compare with the conventional receiving CMUT. FEA simulation results and further discussions of the modified membrane are presented in this section.

4.1 Position configuration

This part will show the optimized position of modified membrane. Firstly, a conventional CMUT receiver is built via FEA to determine the receiving sensitivity. Then a trench with 1 μm width and 0.4 μm height is set in different radial position of the inner part of trench from 11 μm to 20 μm, which means when radial position stretches over 19 μm, the trench width will decrease accordingly and 20 μm represents conventional structure. Receiving sensitivity will be determined as simulation results compared to the conventional CMUT to get the sensitivity enhancement in Fig. 2 (a) and the details from 18 μm to 20 μm is shown in (b). From Fig. 2, it is determined that with the increasing of radial position, better performance is obtained and the optimum value occurs at 18.9 μm, from 19 μm to 20 μm, opposite behavior occurs due to the decreasing of trench width. When the trench is located close to 15 μm, the enhancement is slight. For enhancement consideration, the modified structure should be set on the edge side of the membrane. Even though during the

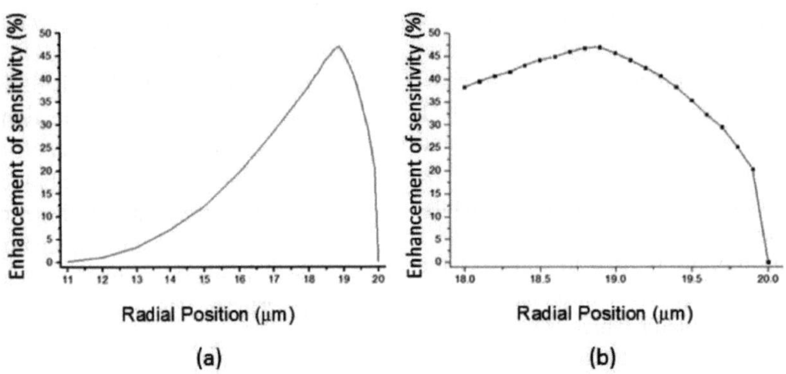

Fig. 2. FEA results of sensitivity enhancement under different radial position for (a) 11 μm to 20 μm with step of 1 μm and (b) 18 μm to 20 μm under step of 0.1 μm.

18 μm to 19 μm region, better performance is presented, however the support part of the membrane part is too thin. It may lead to structural unstable and difficulty in fabricating. Hence in the later simulation, the modified structure is set on edge side start with 18 μm radial position and then gradually closes to mid part of the membrane.

4.2 Slotted membrane

In this slotted membrane part, up to 4 trenches are set surrounding the edge of membrane as ring form as the structure shown in previous Fig. 1 (a). As analyzed in the methodology part, with the modified membrane structure, the collapse voltage will decrease. This section will show FEM result for slotted membrane under 90% collapse voltage. Static average membrane deflection and voltage through the series resistor will be determined as simulation result to calculate coupling coefficient and sensitivity, respectively. For the coupling coefficient calculation, the expression used is shown in equation (5). The simulation results and detailed peak values are displayed in Fig. 3 and Table 3. Simulation results show that with slotted membrane, both coupling coefficient and receiving sensitivity are improved. With the increasing of trenches quantities up to 4, the enhancement increases at most 20.9% and 50.5% of 4 trenches membrane for coupling coefficient and receiving sensitivity respectively. As predicted before, by using the slotted membrane, center frequency will downward shift and more enhancement will bring larger frequency shift. What is more, the bandwidth decreases and it is not predicted before.

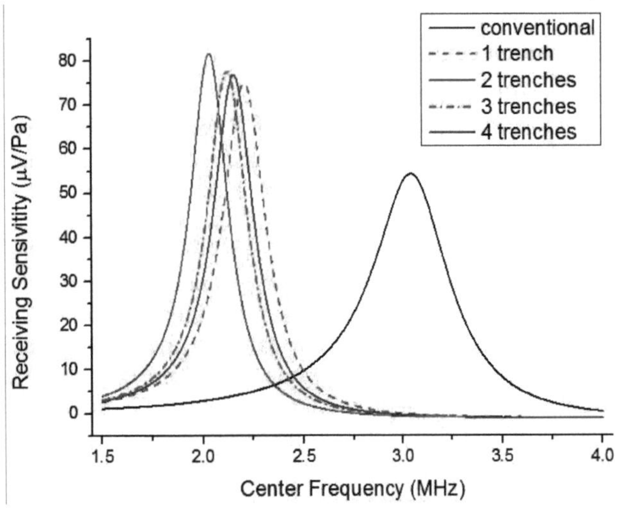

Fig. 3. Comparison of receiving sensitivity of slotted membrane and conventional CMUT. The collapse voltage of conventional CMUT and slotted from 1 to 4 are 184.5 V, 134.7 V, 129.8 V, 127.4 V and 121.9 V respectively.

Table 3. CMUT modeling dimensions in FEM simulation

Quantity of trenches	0	1	2	3	4
Coupling coefficient K_T^2	0.34	0.369	0.371	0.385	0.411
K_T^2 enhancement (%)	0	8.53	9.12	13.2	20.9
Receiving sensitivity (μV/Pa)	54.3	75.1	76.9	78.0	81.7
Sensitivity enhancement (%)	0	38.3	41.6	43.6	50.5
Center frequency (MHz)	3.04	2.21	2.16	2.12	2.03

4.3 Corrugated membrane

This part will present the FEA simulation results for corrugated membrane structure. Up to 3 rings corrugated membrane are implemented in the simulation. Similar to the process of slotted membrane, the simulation results and details peak values of K_T^2 and sensitivity are shown in Fig. 4 and Table 4. With the corrugated membrane, similar to the previous one, K_T^2 and sensitivity are also improved with some degree of frequency shifts. However on the contrary to slotted membrane results, with the increase of rings, the performance will decrease which means 1 ring corrugated membrane has the optimized performance for an enhancement of 5% and 38.3% for K_T^2 and sensitivity respectively. Compared to the slotted membrane, the enhancement of corrugated membrane is less while frequency downward shift also decreases. The reduction of bandwidth also exists after corrugated membrane.

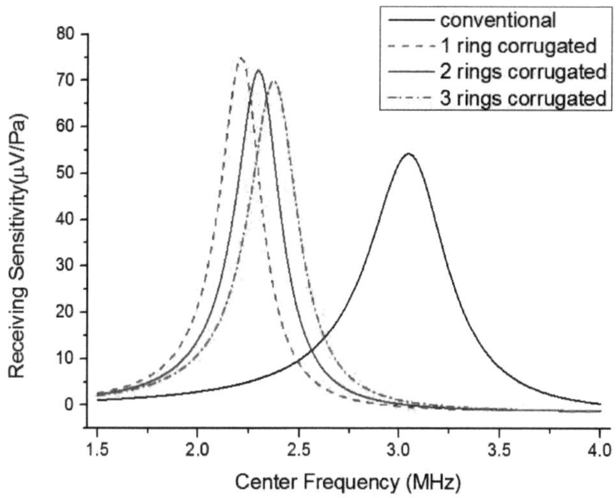

Fig. 4. Comparison of receiving sensitivity of corrugated membrane and conventional CMUT. The collapse voltage of conventional CMUT and corrugated from 1 to 3 are 184.5 V, 132.8 V, 137.7 V and 143.2 V, respectively.

Table 4. CMUT modeling dimensions in FEM simulation

Quantity of corrugated rings	0	1	2	3
Coupling coefficient K_T^2	0.34	0.357	0.353	0.3525
K_T^2 enhancement (%)	0	5.00	3.82	3.68
Receiving sensitivity (μV/Pa)	54.3	75.1	72.2	70.0
Sensitivity enhancement (%)	0	38.3	33.0	28.9
Center frequency (MHz)	3.04	2.21	2.29	2.37

5 Conclusion

This paper presents two improving methods on receiver CMUT by modifying the membrane to the structure of slotted membrane and corrugated membrane. Supported by the methodology analysis and FEA simulation results, both static membrane deflection under electrostatic force of DC bias voltage and harmonic membrane displacement by acoustic pressure are improved with the modified membrane structures. This will lead to the enhancement of electromechanism coupling coefficient and receiving sensitivity respectively. For the slotted membrane, within the range of 1 to 4 trenches, 4 trenches structure has the maximum enhancement. According to the section 4.1 and 4.2, continue adding the quantity of trench may bring more enhancements, however the improvement changes slightly since the radial position is more close to the midpoint of membrane.

Hence 4 trenches structure may be the optimal design for the slotted membrane under the given size and material configurations. For the corrugated membrane, 1 ring structure has the optimized performance. Compare to the corrugated membrane, slotted membrane structure has a better performance on a CMUT receiver, however center frequency will reduce more. For fabrication consideration, the processing of slotted membrane is easier compared with the corrugated membrane.

Since the modified membrane structures are applied on the top surface of the membrane, the interconnection will be affected and 2D asymmetric FEM model cannot describe the situation. Hence 3D FEM model with aluminum interconnection on the surface on modified membrane was simulated. The enhancement of receiving performance and the frequency shift are very close. This represents that the modified membrane structures will also enhance the receiving performance of CMUTs when the interconnection is set in multi CMUT cell fabrication.

Although these modified membrane structures can improve the performance of reception CMUTs, the center frequency downward shift is an issue worth considering which could be recovered by other methods such us thinning membrane. Also solving the reduction of bandwidth and additional fabrication process will be researched in future.

Acknowledgement

The authors would like to express gratitude for the support by the Science and Technology Development Fund of Macau (FDCT) and the Research Committee of the University of Macau. This work was supported by the FDCT under 016/2012/A1,087/2012/A3, 047/2013/A2, 088/2016/A2, and 093/2015/A3 and the University of Macau under Grants MYRG2014- 00010-AMSV, MYRG079(Y1-L2)-FST12-VMI, MYRG 103(Y1-L3)-FST13-VMI.

References

1. Muhammed Sabri Salim, M.F. Abd Male R.B.W. Heng, K.M.Juni, Naseer Sabri Capacitive Micromachined Ultrasonic Transducer: Technology and Application, Journal of Medical Ultrasound (2012) 20, pp 8-31.

2. M. I. Haller and BT. Khuri-Yakub, A surface micromachined electrostatic ultrasonic imaging in air, IEEE Transactions on Ultrasonics, Ferroelectrics, and Frequency Control, Vol. 43, pp. 1-6, Jan. 1996.
3. Zhou, W., Yu, T., Yu, F. A 1-D Theoretical Receiving Model for CMUT. Intelligent Computation Technology and Automation (ICICTA), 2008 International Conference, pp 529-534
4. You, W., Cretu, E., Rohling, R. Improving CMUT receiving sensitivity using parametric amplification. 2011 IEEE International Ultrasonics Symposium, pp 588-591.
5. S. Zhou et al., Improving the Performance of Capacitive Micromachined Ultrasound Transducers using Modified Membrane and Support Structures. In IEEE Ultrasonics symposium, 2005, pp. 1925-1929.
6. Q. Zou, Z. Li, L. Liu Design and Fabrication of silicon condenser microphone using corrugated diaphragm technology, J. Microelectromech. Syst. 5 (1996) 197-204
7. Ladabaum I, Jin X, Soh HT, Atalar A, and Khuri-Yakub BT, ''Surface Micromachined Capacitive Ultrasonic Transducers," IEEE Transactions on Ultrasonics, Ferroelectrics and Frequency Control, vol. 45, May 1998, pp. 678-690
8. Baumgart F. (2000). "Stiffness-an unknown world of mechanical science?". Injury (Elsevier) 31. doi:10.1016/S0020-1383(00)80040-6. Retrieved 2012-05-04.
9. I.O. Wygant, M. Kupnik, B. T. Khuri-Yakub, Analytically calculating membrane displacement and the equivalent circuit ,odel of a circular CMUT cell, Proc. IEEE Ultrasonic Symp., 2008, pp.2111-2114
10. P.R. Scheeper, W. Olthuis, P. Bergveld, The design, fabrication and testing of corrugated silicon nitride diaphragm technology, J. Microelectromech. Syst. 3 (1994) 34-42.
11. Lei, A., Diederichsen, S. E., Hansen, S. M., Stuart, M. B., Bouzari, H., Jensen, J. A., Thomsen, E. V. (2015, 21-24 Oct. 2015). Output pressure and harmonic characteristics of a CMUT as function of bias and excitation voltage. Paper presented at the Ultrasonics Symposium (IUS), 2015 IEEE International.
12. Yaralioglu, G. G., Ergun, A. S., Bayram, B., Haeggstrom, E., Khuri-Yakub, B. T. (2003). Calculation and measurement of electromechanical coupling coefficient of capacitive micromachined ultrasonic transducers. IEEE Transactions on Ultrasonics, Ferroelectrics, and Frequency Control, 50(4), 449-456.
13. Goksen G.Yaralioglu, A,Sanli Ergun, Butrus T.Khuri-Yakub, Finite-Element Analysis of Capacitive Micromachined Ultrasonic Transducers. IEEE Transactions on Ultrasonics, Ferroelectrics, and Frequency Control, Vol.52,No. 12, December 2005,pp.2185-2198

Design of a Silent Speech Interface using Facial Gesture Recognition and Electromyography

Aishwarya Nair, Niranjana Shashikumar, S.Vidhya, and S.K. Kirthika

Department of Sensor and Biomedical Technology, VIT University,Vellore
{aishwarya.nair2012,niranjana.shashikumar2012}@vitalum.ac.in
svidhyavalentina@vit.ac.in, kirthikakumar14@gmail.com

Abstract. A Silent Speech Interface (SSI) aims to substitute natural speech production using various electronic techniques to aid the vocally challenged or impaired. Such interfaces have been experimentally developed using several technologies such as imaging of the lip, tongue and throat regions; mapping of articulatory information, direct mapping of electroencephalogram details, Brain-Computer Interfaces and lip movement tracking mechanisms. Several unimodal as well as multi-modal systems are in the experimental stages. The SSI discussed in this paper was built using lip region segmentation from facial images via Chan-Vese algorithm, simple feature extraction from these images, matching features to words using a Multi-Class Support Vector Machine Classifier and a text - to-speech module producing synthesized speech implemented using MATLAB, with a recognition accuracy of 97.5% for four words – 'cat', 'dog', 'eat' and 'mum'. A multimodal system with facial electromyographic input for validation purposes has been explored.

Keywords: Assistive speech technology · Silent speech interface · Image segmentation · SVM

1 Introduction

The natural mechanism for human communication is voicing speech [1]. However, in several situations – such as in the case of congenital speech defects or after laryngectomy or other surgeries of the vocal tract – human beings are unable to use this method for normal communication. Silent Speech Interfaces are designed to enable speech communication in the absence of pertinent and useful audible signals from the natural speech production pathway. Such devices can be used in enabling speech-impaired individuals to communicate, ranging from basic words like 'on' and 'off' to complex sentences at conversational rates. In events where oral communication cannot be possible, Silent Speech Interfaces stand to be a suitable and viable rehabilitative option. Such interfaces could also potentially be used in situations where discretion is of high importance and in environments of high disturbance where it is difficult to produce an intelligible audio signal. Currently available Silent Speech Interfaces use various methodologies for acquiring and processing signals and designing solutions; used in standalone models as well as multimodal systems. Some examples are devices taking input from brain activity by recording an electroencephalogram, direct reading of neural activity in the formation of words by using Brain Computer Interfaces (BCIs), conversion of sensed articulatory motion into speech using an Electromagnetic Articulograph (EA), detection and extraction of lip contours via various processing algorithms to produce audible speech, vocal tract feature extraction from ultrasound imaging to drive speech synthesis, surface and facial electromyography to create speech signals from the movement of muscles during word formation, etc. Non-invasive systems, including visual processing via optical cameras, are an interesting area of research, with possible refinement via other modes. These methods vary widely in terms of invasiveness, cost, market-readiness, efficiency, ease of usage and success rates of the finished product, but the products have not trickled down into the moderately priced to low-cost end of the economic spectrum. Thus, very few devices actively make it out of laboratories and into the marketspace, which is still dominated by pneumo-larynxes and electro-larynxes as a means for producing silent speech. These devices require some level of vocalisation or airflow in the respiratory pathway to perform their function, and hence cannot be grouped with Silent Speech Interfaces. Yet, they provide consumers with low-cost solutions and are widely used. A recent example is the 'Aum' voice prosthesis developed in Bengaluru, India, which caters to throat cancer patients and is priced at less than USD 1. Among the SSI technologies discussed, special attention is now being given to Brain Computer Interfaces (BCIs) and multimodal systems, from a research perspective. BCIs are useful for individuals who no longer possess control of their muscular activity. This field is being researched for various potential uses, especially Artificial Intelligence (AI). Advances will most likely be seen in the next 5 years.

Multimodal systems offer the benefit of accuracy via redundancy. Among SSIs, this is a regular phenomenon, an example being integrating ultrasound and optical input of the jaw region. These systems are especially useful in creat-

J. Goh et al. (eds.), *The 16th International Conference on Biomedical Engineering*,
IFMBE Proceedings 61,
DOI: 10.1007/978-981-10-4220-1_22

ing customised solutions catering to the comfort and capabilities of the user, but increasing complexity leads to solutions that are difficult to maintain and troubleshoot.

The aim of this paper is to explore a user-friendly, non-invasive communication device for speech-impaired individuals – a system known as Silent Speech Interface. The described model will focus on predicting the words mouthed by the user using facial gesture recognition and various classification techniques. A refinement of the model by including facial electromyography is also proposed.

2 Methodology

The proposed solution works on digital lip-reading principles [2], wherein the users position themselves in front of the camera and mouth a word without any audio input, and in turn, hear the word in the form of computerized speech.

The process is initiated by a Graphical User Interface (GUI), which performs processing on a training database, which consists of 20 samples each of different people mouthing four different words – 'hello', 'cat', 'dog' and 'eat'.

The final methodology employed was as follows:

a) A video of the user saying one of the four words is acquired and is displayed in the GUI.
b) The video is split into frames and the Chan-Vese Segmentation algorithm is applied to all relevant frames.
c) Features such as Area, Convex Area, Equivalent Diameter, Eccentricity, Orientation, Major Axis length and Minor Axis length are extracted from all segmented frames.
d) The Fine Gaussian SVM classifier, which has been trained with similar data, is provided with the extracted features, based on which it classifies the test video to one of the four classes.
e) The identified class label (word) is then displayed as text in the GUI.
f) The Text-to-Speech module converts the displayed text to voice output.

2.1 Data Collection

To validate the segmentation procedure and obtain sufficient information to observe patterns that exist in the data, a collection of video sequences was required. Creation of a sufficiently large database is complex and time consuming with several ethical conundrums. Thus, the database consists of four students of VIT University, Vellore uttering the words 'cat', 'dog', 'eat' and 'hello' five times each. Each time a word was uttered, a video was captured and its runtime was truncated to a value between one and two seconds to maintain a standard. The subjects were requested to wear black lipstick to increase the visibility. The video camera was positioned such that only the region of face below the nose bridge was captured. The segmentation algorithm detailed in the next section employs a colour to greyscale transformation and this causes the lip region to become indistinguishable from the background. Many image enhancement techniques were applied at this stage to accentuate the lip region, but were not as effective at aiding the segmentation as the application of black lipstick before video capture. The videos were the split into frames as detailed in Chapter 3. Features were extracted from each frame and tabulated, following which they were fed to the classifier for training.

2.2 Segmentation

Over the years, detection of the lip contours has been employed in gesture recognition systems where the recognized contours are used to drive many other applications. The Chan-Vese model [3] is a powerful and flexible implementation of the active contours model. The initial contour is

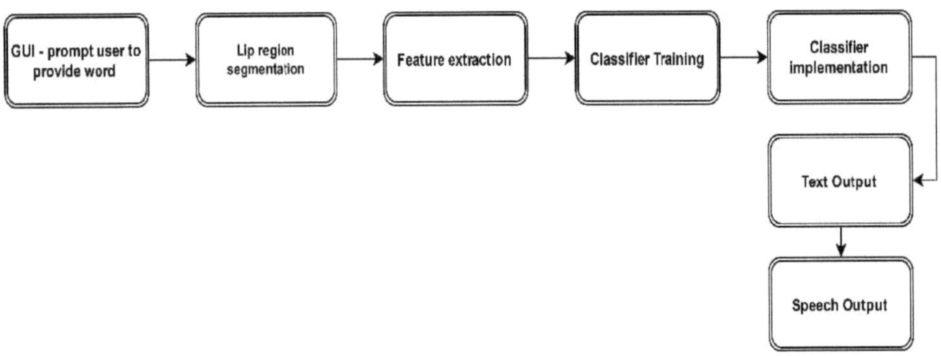

Fig. 1. Functional Block Diagram of the methodology

to be specified by the user and then the curve deforms as per an evolution equation, which is also specified by the user. The evolution equation can be formulated via several methods. The energy functional which comprises of both internal and external energy terms generally has several parameters and searching for a solution that minimizes this functional can be a complex computational task. Instead of trying the optimal solution of the curve that minimizes the energy functional, it may be easier to redefine the problem in terms of level sets. This is the basis of the Chan-Vese algorithm. In this method, a function is defined with the following parameters: image co-ordinates in the x and y directions and time *t*. This is called the level set function. For any instant of time *t*, the level set provides a boundary contour and an image portion segmented based on the provided boundary. When this segmented image is validated and it is determined that the segmentation is imperfect, the level set evolves and the procedure is repeated until the stopping point is reached.

Area	*Convex Area*	*Equivalent Diameter*	*Eccent-ricity*	*Orientatio-n*	*Major Axis*	*Minor Axis*
36849	38300	216.6047	0.545	-83.7716	239.51	200.67

The Chan-Vese Segmentation algorithm was applied to selected frames of each video sample in the collected database and features were extracted which were later used in the training of the classifier. Information obtained from the segmentation process is tabulated as seen in Table 1. The values obtained are in accordance with the default measurement scheme employed by MATLAB's inbuilt command 'regionprops'.

2.3 Classifier Training and Implementation

A variety of classifiers was evaluated for the proposed SSI. The minimum requirements were capability of employing supervised learning algorithms, quick training process and easy deployment. Neural networks do not produce consistent results when the quantum of data is relatively low. Linear Discriminant Analysis classifiers provided low accuracy for project data, and k-NN classifiers did not learn from the training data, instead using the data from scratch for each iteration. Decision trees tend to develop overly complicated trees that do not deal well with generalization of data, and thus were rejected in favor of Support Vector Machines (SVMs).

SVMs are machine learning models that allow for supervised learning with dedicated algorithms and analysis techniques that are used to solve classification problems. When provided with labelled training data, an SVM algorithm attempts to create a classification model that can classify similar but new data into either class. Traditional SVMs

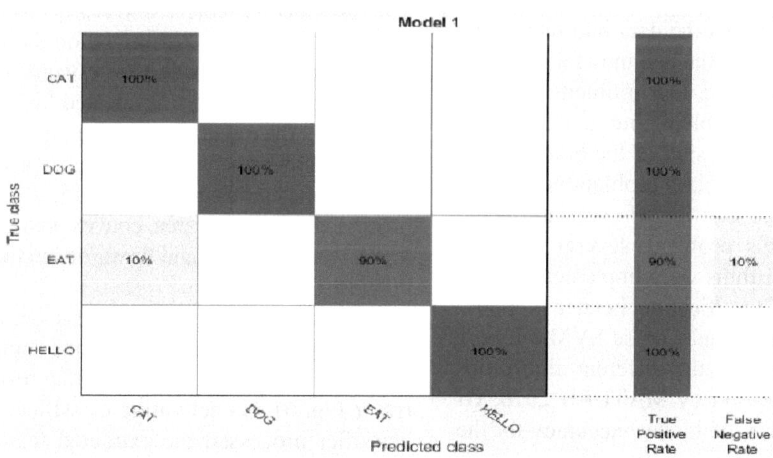

Fig. 2. Classifier training – confusion matrix, illustrating True Class vs Predicted Class performance for each of the words.

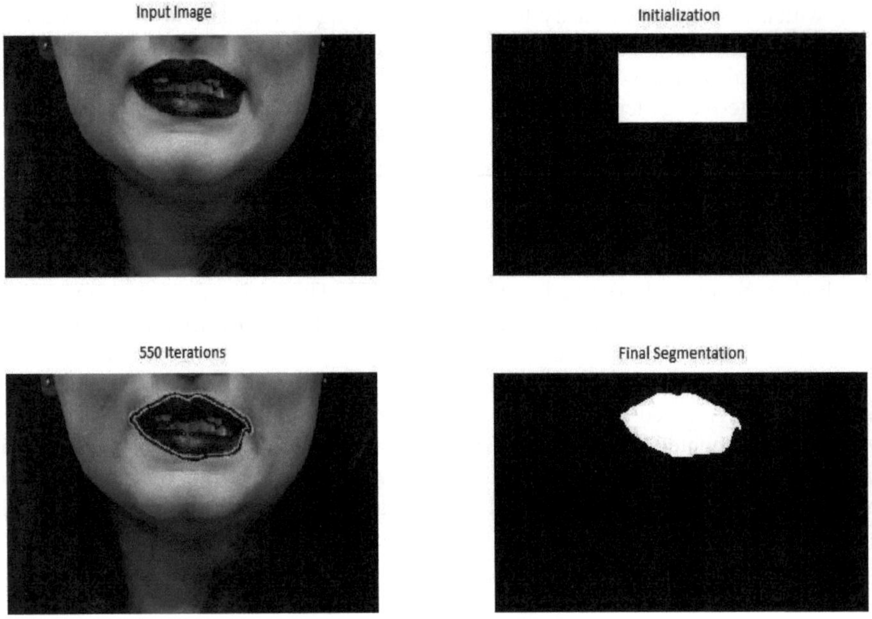

Fig. 3. Segmentation of a single frame

are binary and can only categorize data into one of two classes. A simple illustration of how the SVM algorithms work can be achieved by visualizing the training data as points in space. The SVM classifier attempts to categorize the examples belonging to each group while maintaining a clear divide in space between the two groups. The wider this gap is, the better the classification. When the classifier is presented with new data, it allots the data into one of the two classes based on which side of the gap they fall on. This gap is often called a hyperplane [4]. Our problem however is not a binary classification problem, but a multi-class problem. For this purpose, an extension of the binary SVM was considered in which the multiclass problem is reduced to binary problem. To validate that SVM's are indeed the best possible classifier for this problem, several types of Support Vector Machine algorithms were implemented on a trail run via the MATLAB Classification Leaner Application. Cubic SVM, Fine, Medium and Coarse SVMs, Linear and Quadratic SVMs are some of the different algorithms provided by the application hosted by MATLAB 2016. Of these, Fine Gaussian SVM gave the highest accuracy for the training data.

The overall accuracy of the trained classifier was 97.5%, with 10% error in True Class vs Predicted Class confusion matrix for the word 'eat' (Fig. 2.). Intuitively, this seems plausible as the words 'eat' and 'cat' may have similar initial area values as both words involve significant opening of the mouth.

3 Results

The test video is subjected to segmentation and feature extraction. Segmentation is performed on the frames with the maximum relevant information for a word, for 550 or more iterations, to get a perfectly segmented binary image of the lip region, while the mouth is in motion. As shown in Fig. 4 and Fig. 5, every fifth frame was considered in a video captured at 25fps (thus, frame numbers 1, 5, 10, 15, 20, 25 were selected and processed). Features (area, convex area etc.) for each processed frame were extracted and forwarded to the classifier.

The above segmentation process begins with a GUI that plays the test video, and performs segmentation and classification on the activation of the respective push-buttons (refer Fig. 6). On activating classification, the trained SVM classifier processed the extracted features, and assigns the video to a specific class. This class (word) is then displayed in the GUI as the recognised word. Electronic/computerised speech is used via a text-to-speech interfacing module to generate audio.

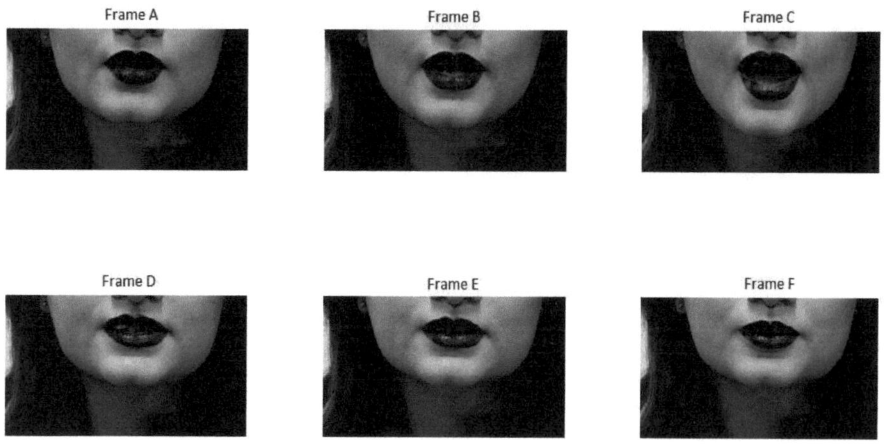

Fig. 4. Multiple frames of subject pronouncing a single word

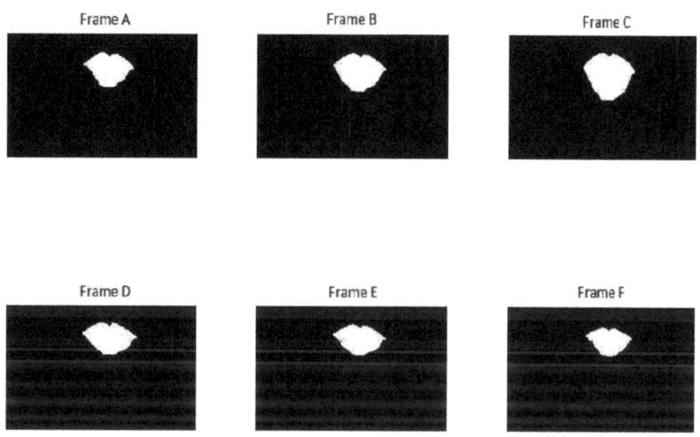

Fig. 5. Final segmentation of the corresponding frames shown in the previous figure

4 Discussion

The proposed solution is easy to learn, easy-to-use, technology-driven and accurate. The device is to act as a suitable substitute for voiced speech and/or as an aid for the vocally challenged or impaired. Alternatively, it can be used as a communication tool in noisy environments.

Future work includes incorporation of laryngeal EMG which serves to differentiate between optically similar words. For example, the words DOG and CLOCK have quite similar mouth movements and can be misjudged by the designed purely visual system. The difference between the words is evident when one observes the tracheal muscle movement when the word is vocalized. Contrary to popular opinion, it is possible to view this movement even when the person is unable to produce the sound. However, this view is not currently supported by significant research and in depth studies will need to be carried out to validate this hypothesis.

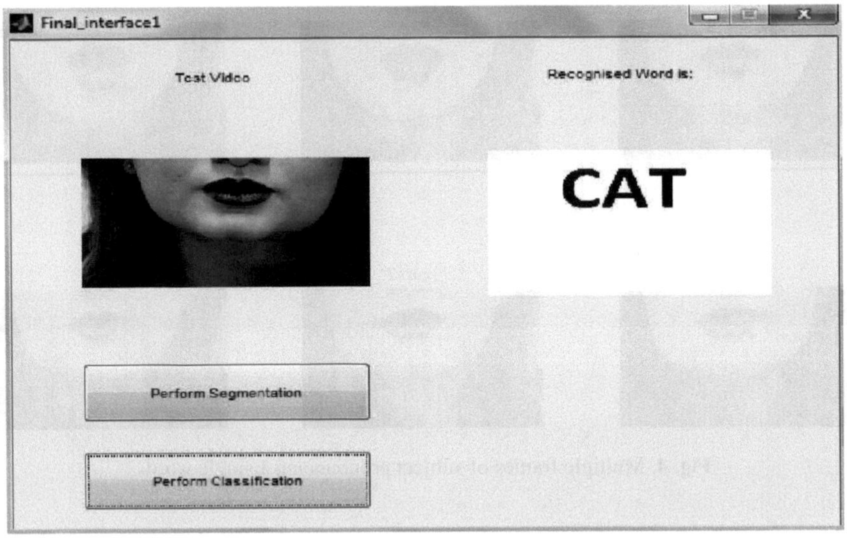

Fig. 6. GUI displaying the corresponding class-label (word)

Silent Speech Interfaces are currently in experimental stages and these devices have not commercially reached the assistive devices market. The reasons for this have been outlined in the introduction to this paper. Most SSIs today are tethered to large systems that do real time processing of the data, which is of course the desired result, but such systems are harder to commercialize because one of the trade-offs would be the ability to process data in real time versus portability. The solution proposed in this paper circumvents this issue by not attempting to process the data in real time. If a significantly large database of words can be obtained with multiple speakers with varying facial features, then the classifier can be taught to classify more words with greater efficiency. This type of an approach to silent speech devices offers the possibility of integration into mobile systems like smart phones and PDA's. Considering the portability of computing languages and the relative ease with which mobile applications are produced, it can be safely assumed that silent speech devices hosted on a mobile environment can become a reality sooner than expected.

References

1. Denby, B., Oussar, Y., Dreyfus, G., Stone, M.: Prospects for a Silent Speech Interface Using Ultrasound Imaging. IEEE International Conference on Acoustics, Speech and Signal Processing, Toulouse, France, 1365–1368 (2006)
2. Nakamura, H.: Method of recognizing speech using a lip image. United States Patent 4769845 (1988)
3. Chan, T. F., Vese, L., A.: Active contours without edges. IEEE Transactions on Image Processing, vol. 10(2), pp. 266-277 (2001)
4. 4. Kotsiantis, S.B.: Supervised Machine Learning: A Review of Classification Techniques. Informatica, vol. 31, pp. 249-268 (2007)

Performance Comparison of Oral Cancer Classification with Gaussian Mixture Measures and Multi Layer Perceptron

Harikumar Rajaguru and Sunil Kumar Prabhakar

Department of ECE, Bannari Amman Institute of Technology, India

Abstract. One of the most commonly occurring cancers is oral cancer. The incidence of the oral cancer seems to be increasing exponentially in the world. The clinician has to undergo a higher level of dilemma every time in order to differentiate the cancerous lesions from other controversial and poorly defined lesions that are present in the oral cavity. Early stage carcinomas and its subsequent manifestations are highly misinterpreted because at the initial stage there is minimum discomfort in the patient and they simply mimic many similar benign lesions. The analysis to be done by the doctors is often delayed and therefore there is a high risk for the cancer to spread in the body. Squamous cell carcinoma is the most common malignant neoplasm present in the oral cavity. Therefore the accurate diagnosis and management of this particular Squamous cell carcinoma which originates from the surface of the oral muscle has to be done well. The main aim of this work is to assess the clinical features, diagnostic procedures and treatment required for oral cancer patients. The staging of the cancer is generally divided into two stages namely, clinical and pathological. In TNM (Tumour, Node, Metasis), a lot of novel prognostic tools have been traced and new methodologies for the prognostic factors have been drastically improved and developed. This paper compares the classification accuracy of the TNM staging system with the aid of Multi Layer Perceptron (MLP) and Gaussian Mixture Model (GMM) classifiers. In this work, totally 75 oral cancer patients are studied. For both the classifiers, the input variables are nothing but the TNM variables such as tumour size, number of positive regional nodes, distance metastasis, hereditary etc. Out of the two post classifiers utilized here, GMM provided a better result as of 94.18% average accuracy for all the stages while Multi Layer Perceptron (MLP) showed an average accuracy of about 89.5% for all the stages. In this paper, Extreme Learning Machines (ELM) is also employed as a post classifier later for the oral cancer classification and the performance of it is compared to the performance of both the GMM and MLP.

Keywords: Carcinoma, TNM, GMM, MLP, ELM, Accuracy

1 Introduction

In majority of the population, one of the commonly occurring diseases is oral cancer [1]. Diagnosing it medically is extremely crucial and an intricate task to perform because it requires a high amount of proficiency and exactness. Oral cancer begins in the oral cavity which includes the entire mouth, lips, the inner lining of the lips and cheeks, the teeth gums, the portion of the mouth below the tongue, the hard palate or the bony roof of the mouth, the area behind the wisdom teeth and etc [2]. If oral cancer can occur, it can easily spread to various other parts such as lungs and neck. In the lymph nodes of the neck, a common metastasis occurs so that the cancer cells spread throughout. Due to the drastic and rigorous improvements in the treatment of oral cancer, a high rate of oral cancer disappearance has been achieved, but there are high chances for the disease to evolve again, that is, remission stage can be easily attained. At the starting stage of the remission process, no clinical and laboratory evidences can be found out and hence the patient is considered only cancer free. But suddenly at any point of time some particles of diseases which are invisible to eyes might emerge out which leads to the immediate spread of this deadly disease. In the context of oral cancer, the occurrence of the remission stage is pretty high and therefore one has to maintain a very healthy lifestyle. The patient who suffer from oral cavity cancer has to cope up with the terrible effect of this disease, such that the lack of ability to eat well and speak well, physical appearance of the face due to the surgical procedures and treatment and so on. A lot of factors attribute to the reoccurrence of cancer which involves factors such as infection site, age, sex, primary tumour location and its respective histological features etc.

The main symptoms of oral cavity cancer are severe pain in the mouth that does not heal for months, a lump in facial cheeks, problems in chewing and swallowing, movement of the tongue and jaw becomes difficult, loosening of the teeth, severe pain in the teeth accompanied by voice changes, severe weight loss, prolonged bad breadth, red patches on the tonsils, gums and tongue [3]. If the above symptoms persist for more than two weeks, it is recommended to

J. Goh et al. (eds.), *The 16th International Conference on Biomedical Engineering*,
IFMBE Proceedings 61,
DOI: 10.1007/978-981-10-4220-1_23

consult an oncologist as early as possible. Recently, with the advent of the soft computing techniques, the classification of oral cavity cancer can be done reliably. Soft computing is a particular or special branch of computational intelligence unit which employs a lot of statistical, optimization and probabilistic measures that allows the computer to learn from the previous examples and to trace the hardest patterns from very large, complex and noisy data sets. This feature of soft computing techniques suits well for medical applications especially, those dependent on genomic and complex proteomic measurements. So computational intelligence technique is widely used for cancer detection, prediction and prognosis. A lot of work has been proposed recently in literature for the oral cancer classification and some important work is reviewed in this paper. Parkin et al showed the estimates of the worldwide incidence of twenty five cancers [4]. Aziz showed the oral sub mucous fibrosis as an unusual disease [5]. A national epidemiological survey of oral mucosal lesions was reported by Zain et al [6]. The application of artificial intelligence to identify people at risk of oral cancer was done by Rosmai et al [7]. The prediction of oral cancer survivability using classification algorithms was performed by Kaladhar et al [8]. The determination of the critical success factors of oral cancer susceptibility prediction using fuzzy models was developed by Rosmai et al [9]. The prediction of Hospital charges for cancer patients with the help of data mining techniques was done by Kang et al [10]. A framework for early detection and prevention of oral cancer using data mining was reported by Sharma and Om [11]. The advances in oral cancer detection using optical coherence Tomography was reported by W.Jung et.al [12]. The diagnosis of oral cancer using genetic programming was done by Simon Kent [13]. The oral lesion classification using true colour image was done by Chodorowski et al [14]. The wavelet based texture classification of oral histopathological sections was done by M.R.Krishnan et al [15]. The performance of data mining technique for oral cancer prediction was compared by Neha Sharma et al [16]. Colour-based tumor segmentation for the automated estimation of oral cancer parameters was done by Yung-nien Sun [17]. A novel wavelet neural network based pathological stage detection technique for oral precancerous condition was represented by R.R.Paul et al [18]. In this work, the usage of post classifiers like GMM, MLP and ELM were employed to classify the oral cancer levels. The block diagram of the work is shown in Figure 1.

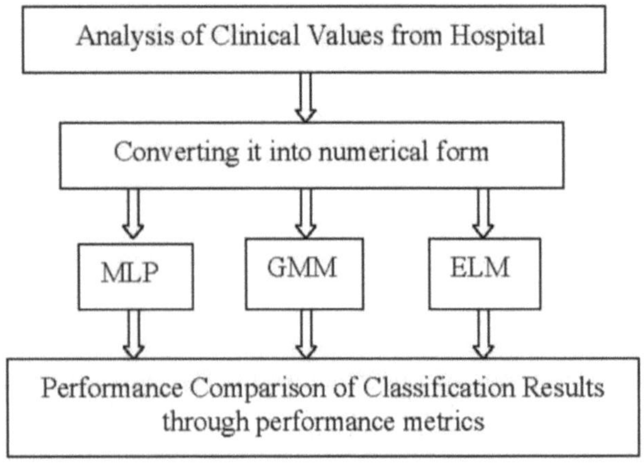

Figure 1 Block Diagram of the Classification Procedure

2 Materials and Methods

For a total of 75 patients who suffer from oral cancer, the dataset was obtained from Department of Oncology of G Kuppuswamy Naidu Hospital (GKNM) Hospital, Coimbatore, India. Based on the reports from referral letters, operative reports, hospital charts and radiological studies, radiation therapy and pathological report, this study is mainly dependent on it. Based on the various factors such as tobacco chewing habit, smoking, quality of life style, food habits, tumour size and location, sex and age, the patients are analyzed. Based on the International Union against cancer (UICC), the tumour is clinically staged into 4 stages such as stage I, stage II, stage III and stage IV. Stage I has 16 patients, stage II has 22 patients, stage III has 26 patients and stage IV has 11 patients.

3 Classifiers used here

After the patients are grouped in the four stages, the post classifiers such as MLP, GMM and ELM are employed to analyze the oral cancer here.

3.1 MLP

The multilayer perceptron neural network is a feed forward type of neural network which has one or more hidden layers [19]. MLP networks with one or more hidden layers have a good capability for approximating the continuous functions upto a certain level of accuracy. Depending on the structure of the neural network and the training algorithm applied, the performance of classification of the MLP is determined. The training algorithms called Levenberg-Marquardt (LM) is utilized to examine the applicability of the MLP network. MLP can apply non linear discriminant for classification purposes and non linear regression functions for regression. MLP is usually called or two-layer network as there is no computation performed in the input layer. The non linear transformation is made by the hidden units from the d dimensional input space to the H dimensional space which is spanned by the hidden units. In the obtained new H dimensional space, the linear function is implemented by the output layer. For implementing more complex function of the inputs, multiple hidden layers are used often but to learn the weights in the network in such deep networks is a more complicated task.

LM Algorithm

As the LM algorithm has a better learning rate and a relatively high stability when compared to the famous Back Propagation Algorithm, it is used here. The Gauss-Newton algorithm can be approximated to get a LM algorithm, so that a faster learning rate is obtained than the BP algorithm which uses steepest descent algorithm. The LM modification to the standard Gauss-Newton algorithm is expressed as follows

$$\Delta x = \left[J^T(x)J(x) + \mu I\right]^{-1} J^T(x)e(x)$$

where μ represents the Marquardt adjustment parameters and I represents the identity matrix. If μ is small, then the approximation of Gauss-Newton algorithm is done by LM algorithm. Since the convergence is faster and more accurate in Gauss-Newton algorithm, shifting it quickly is possible. Unless the error change obtained is positive, the value of μ is decreased. If the value of μ is large, then the LM algorithm is used to approximate the steepest descent algorithm.

The standard results of the MLP back propagation neural models when trained with the LM learning algorithm are shown in Table 1. The momentum α (0.5), learning rate or gain rate η (0.3) and various training epochs are tabulated for each MLP architecture

Table 1 Estimation of Mean Square Error in Various MLP Network Architectures

Architecture	Training Epochs	Mean Square Error (MSE) Index	
		Training	Testing
24-12-4	38	0	7.31E-03
16-3-1	6	0	2.19E-02
8-8-1	283	0	9.13E-03
8-4-1	6	0	5.1E-02
4-4-1	9	0	2.83E-08
4-4-4	12	0	7.74E-03
2-2-2	3820	3.0E-08	3.7 E-08
2-4-2	8	0	0
1-1-1	4538	1.08E-08	1.2E-08

In this case the architectures selected are (24-12-4), (4-4-1) and (2-4-2) MLP network architectures because they have lesser number of training epochs. If more number of training epochs are considered then it means that the system can be easily ruined due to the high complexity.

3.2 Gaussian Mixture Models

The finite mixture models always provide a very flexible approach for the purposes of statistical modeling of a wide range of random phenomena. The flexible approach is because it has the capability to trap the heterogeneity in the context of cluster analysis. A finite mixture of normal densities can approximate any continuous distribution very well and so GMM provides a good semi parametric structure and shape [20]. When a single parametric family does not give a satisfactory model for observing the local variations in the data, then mixtures can be used. Also the GMM should provide the scope of inference in a versatile manner. Let $(A_1,\ldots,A_n)$ be assumed as a random sample of size 'n' where Aj is nothing but a 'd'-dimensional random vector with probability distribution $f(a_j)$ on Q^d. The vector of the nodal parameters is denoted as θ. The g-component mixture models in a suitable parametric form is given as

$$f(a_j;\psi) = \sum_{i=1}^{g} \pi_i f_i(a_j,\theta_i)$$

where $\psi = (\pi_1,\ldots,\pi_g,\theta_1,\ldots,\theta_g)^T$ has the unknown parameters. The model parameters for the ith mixture component is represented as θ_i and $\pi = (\pi_1,\ldots,\pi_g)^T$ denotes the vector values of the mixing properties with $\sum_{i=1}^{g}\pi_i = 1$. The mixture components has a multivariate Gaussian distribution and each component in the mixture is represented by

$$f(a_j;\theta_i) = (2\pi)^{-1}|\Sigma_i|^{-1/2}\exp\left\{-\frac{1}{2}(a_j-\mu_i)^T\Sigma_i^{-1}(a_j-\mu_i)\right\} = \phi(a_j;\mu_i,\Sigma_i)$$

where the multivariate Gaussian density is denoted by ϕ with its respective mean vector μ_i and a covariance matrix $\Sigma_i, i = 1,\ldots,g$.

Ultimately the Gaussian mixture model is of the form

$$f(a_j;\psi) = \sum_{i=1}^{g} \pi_i \phi(a_j.\mu_i,\Sigma_i)$$

For the evaluation of the mixture distributions, techniques such as graphical models, maximum likelihood (based on EM), method of moments, and Bayesian approaches are present. In this paper, the most popularly used maximum likelihood technique based on EM is used.

3.3 Extreme Learning Machines (ELM)

Assume the training data which has number of samples is represented by $\{a_i, b_i\}$, where $a_i \in R^p$ and $b_i \in R^q$, a feed forward neural network which has a standard single hidden layer. This type of neural network has hidden neurons an an activation function and $d(a)$ is represented as

$$\sum_{i=1}^{H} \alpha_i d_i(a_j) = \sum_{i=1}^{H} \alpha_i d(w_i.a_j + c_i) = z_j \text{ where } j = 1,\dots,F$$

Here w_i and α_i are known as the weight vectors which connects the various inputs to the ith hidden neurons and its respective output neurons too. For the ith hidden neurons, c_i is denoted as the threshold and z_j is denoted as the output from a single hidden layer feed forward neural network for the particular data point j. The random generation of the weight vector w_i is done and it is based on a continuous probability distribution. To approximate F training data with zero errors, a standard single layer feed forward neural network with totally H hidden neurons and an activation function $d(a)$ is represented as

$$\sum_{j=1}^{F} \|z_j - o_j\| = 0$$

Therefore $\sum_{i=1}^{H} \alpha_i d(w_i.a_j + c_i) = o_j$, where $j = 1,\dots,F$ for particular values of α_i, w_i, and c_i. The above equation can be expressed in a compact form as follows

$$J\alpha = V$$

where J represents the hidden layer output matrix of the neural network and is expressed as

$$J(w_1,\dots,w_H,c_1,\dots,c_H,a_1,\dots,a_H) = \begin{bmatrix} d(w_1.a_1 + c_1) \dots\dots\dots d(w_H.a_1 + c_H) \\ \vdots \qquad\qquad \vdots \\ d(w_1.a_F + c_1) \dots\dots\dots d(w_H.a_F + c_H) \end{bmatrix}_{F\times H}$$

If the hidden layer output matrix of the neural network J is not known , then a gradient based learning algorithm is utilized so that $\|J\alpha - V\|$ is minimized by adjusting a vector W which is nothing but a set of biases (c_j) and weights (w_j, α_i). It is done iteratively as per the following equation

$$W_p = W_{p-1} - \eta \frac{\partial C(W)}{\partial W}$$

where the learning rate is denoted as and to compute the gradient in a feed forward neural network, Back propagation learning algorithm is utilized well.

If randomly assigned input weights and hidden layer biases are present in the single layer feed forward neural network with any non zero activation function, the universal approximation of any continuous function on any particular input dataset is done. If the least square solution of the linear system is found out, then this problem is solved and represented as

$$\|J(w_1,\dots,w_H,c_1,\dots,c_H)\alpha' - V\| = \min_{\alpha} \|J(w_1,\dots,w_H,c_1,\dots c_H)\alpha - V\|$$

Using the smallest norm least squares solution of $J\alpha = V$, the solution becomes

$$\alpha' = J^{@}V$$

where $J^{@}$ is denoted as the Moore-Penrose generalized inverse of matrix J. The most important property of this solution is that the minimum training error is achieved easily and the best generalization performance is obtained. The Extreme Learning Machine (ELM) as proposed in [21] is summarized as follows:

If a training data set $\{a_i, b_i\}$ is assumed initially, $a_i \in R^n$, $b_i \in R^m$, having totally F number of samples, a standard single hidden layer feed forward neural network algorithm which has H hidden neurons and activation functions $d(x)$ is working as per the following steps.

Step 1: The input weight is randomly assigned along with the bias c_i, $i = 1,\dots.H$.

Step 2: The hidden layer output matrix J is calculated J

Step 3: The output weights α is calculated by $\alpha = J^{@}V$

Step 4: The different activation functions handled in ELM technique are Sigmoid, Radbas and Hard Lim. For 20 hidden neurons and 30 hidden neurons, the analysis is performed and comparison is done.

3.4 Effect of Overfitting in the Classification Procedure:

In the context of statistics and machine learning theory, one of the most common tasks is to fit a "model" to a specific set of training data, so as to be able to make reliable predictions on general untrained data. In over fitting, a statistical model explains random error or noise instead of the underlying relationship. Over fitting can occur when a model is too complex, such as having too many parameters relative to the total number of observations. A model that has been over fit has very poor predictive performance, as it overreacts to even minor fluctuations in the training data. But in our case, though the data base seems to be complex with many parameters, it cannot be treated to the total number of observations because it is staged into 4 different stages of oral cancer based on the severity levels and so the complexity of the data is greatly reduced and there are no chances for the over fitting to occur. As the data is clinically staged into 4 stages such as stage I, stage II, stage III and stage IV. Stage I has 16 patients, stage II has 22 patients, stage III has 26 patients and stage IV has 11 patients. As it is grouped into 4 different stages of oral cancer, the relative number of patients in a particular class or group is quite small and therefore the effect of over fitting is very less for our case and so using three different classifiers such as Gaussian Mixture Model (GMM), Multilayer Perceptron (MLP) and Extreme Learning Machine (ELM), the classification has been done very well and it is proved through the results obtained in the performance analysis and comparison. So the models are definitely not over fitted for the given data set and so the results obtained are quite good and versatile.

4 Results and Discussion

For the crystal clear analysis of the three classifiers , MLP, Elman and GMM for Classification of Oral Cancer, the standardized parameters taken for the analysis is False Alarm measures, Perfect Classification Measures, Missed Classification Measures, sensitivity and specificity measures and accuracy measures are given by the following mathematical formulae as

$$PI = \left(\frac{PC - MC - FA}{PC} \right) \times 100$$

PC is expressed as Perfect Classification, MC is expressed as Missed Classification and FA is expressed as False Alarm. The Sensitivity, Specificity and Accuracy measures are mathematically given by the following formulae

$$Sensitivity = \frac{PC}{PC + FA} \times 100$$

$$Specificity = \frac{PC}{PC + MC} \times 100$$

$$Accuracy = \frac{Sensitivity + Specificity}{2} \times 100$$

The Table 2 shows the results of GMM Classifier. The results of ELM Classifier with Sigmoid Function and 20 hidden neurons are shown in Table 3. The results of ELM Classifier with Hard Lim Function and 20 hidden neurons are shown in Table 4. The Table 5 shows the Results of ELM Classifier with Radbas Function and 20 hidden neurons. Table 6 shows the Results of ELM Classifier with different activation functions for 20 hidden neurons. The Table 7 Results of ELM Classifier with different activation functions for 30 hidden neurons. The Table 8 Consolidated Performance Analysis of all the Post classifiers for Oral cancer classification.

Table 2 Results of GMM Classifier

	PC	MC	FA	PI	Sensitivity	Specificity	Average
T1	100	0	0	100	100	100	100
T2	100	0	0	100	100	100	100
T3	73.68	5.2	21.12	64.27796	77.72152	93.40771	85.56461
T4	82.35	17.65	0	78.56709	100	82.35	91.175
Average	89.0075	5.7125	5.28	85.71126	94.43038	93.93943	94.1849

Table 3 Results of ELM Classifier with Sigmoid Function and 20 hidden neurons

	PC	MC	FA	PI	Sensitivity	Specificity	Average
T1	100	0	0	100	100	100	100
T2	85.12	7.41	7.47	84.94	85.12	85.12	85.12
T3	68.42	10.52	21.06	67.95	76.46	86.67	81.56
T4	76.48	23.52	0	76.17	100	76.48	88.24
Average	82.505	10.3625	7.1325	82.265	90.395	87.0675	88.73

Table 4 Results of ELM Classifier with Hard Lim Function and 20 hidden neurons

	PC	MC	FA	PI	Sensitivity	Specificity	Average
T1	100	0	0	100	100	100	100
T2	96.29	0	3.71	96.25	96.29	100	98.15
T3	94.73	0	5.27	94.67	94.73	100	97.37
T4	88.24	11.76	0	88.11	100	88.24	94.12
Average	94.815	2.94	2.245	94.7575	97.755	97.06	97.41

Table 5 Results of ELM Classifier with Radbas Function and 20 hidden neurons

	PC	MC	FA	PI	Sensitivity	Specificity	Average
T1	83.33	0	16.67	83.12	83.33	100	91.66
T2	85.18	0	14.82	85.1	85.18	100	92.59
T3	63.2	5.3	31.5	62.62	66.73	92.26	79.49
T4	70.58	29.42	0	70.16	100	70.58	85.29
Average	75.5725	8.68	15.7475	75.25	83.81	90.71	87.2575

Table 6 Consolidated Results of ELM Classifier with different activation functions for 20 hidden neurons

Hidden neurons	Activation Function	PC	MC	FA	PI	Sensitivity	Specificity	Average
20	Sigmoid	82.505	10.3625	7.1325	82.265	90.395	87.0675	88.73
	Hard Lim	94.815	2.94	2.245	94.7575	97.755	97.06	97.41
	Radbas	75.5725	8.68	15.7475	75.25	83.81	90.71	87.2575
Average Values		84.2975	7.3275	8.375	84.0908	90.65333	91.6125	91.1325

Table 7 Consolidated Results of ELM Classifier with different activation functions for 30 hidden neurons

Hidden neurons	Activation Function	PC	MC	FA	PI	Sensitivity	Specificity	Average
30	Sigmoid	93.68	1.45	4.87	93.605	93.68	100	96.84
	Hard Lim	78.015	11.66	10.325	66.9425	88.15	86.605	87.38
	Radbas	68.0775	14.4675	17.455	49.985	80.42	83.2325	81.82625
Average		79.92417	9.1925	10.88333	70.1775	87.41667	89.94583	88.68208

Table 8 Consolidated Performance Analysis of all the Post classifiers for Oral cancer classification

S.No	Oral Cancer Stage	% of Classification By Clinical Procedure	% of Classification by MLP	% of Classification by GMM	% of Classification by Elman (Hard Lim Activation Function with 20 neurons
1	T1	98	92	100	100
2	T2	100	89	100	98.15
3	T3	97	87	85.56461	97.37
4	T4	100	90	91.175	94.12

5 Conclusion

Thus oral cancer is one of the significant problems of the health affecting both men and women throughout the world. It is the fourth commonly occurring cancer in men and eighth commonly occurring cancer in women. Tumours arising from odontogenic tissues (tooth forming tissues) are referred as odontogenic tumours and it can be either malignant or benign. Malignant tissues are mostly cancerous as it affects the entire oral cavity which includes tongue, lips, gum tissues and the inner lining of cheeks. The treatment cannot be successful if the oral cancers are identified and diagnosed at a later stage and hence early detection and perfect classification techniques are required. This cancer classification procedure helps the surgeons to come up with novel therapeutic measures for the welfare of the patients. In this paper, three different post classifiers like MLP, GMM and ELM are employed to classify the risk of oral cancers in the different stages. In the T1 and T2 cancer stages, the GMM classifiers provided an accuracy of 100%, in the T3 cancer stage, the ELM classifier with Hardlim activation function with twenty hidden neurons provides an accuracy of 97.37% and in T4 stage, the ELM Classifier with Hardlim activation function with twenty hidden neurons provides an accuracy of 94.12%. All the post classifiers classified the oral cancer with a pretty good percentage of accuracy as shown in the comparison Tables 2-8. Future work plans to implement different schemes of post classifiers for a better classification of oral cancer.

References

[1] R.Harikumar, N.S.Vasanthi, Performance Analysis of Artificial Neural Networks in Classification of Oral Cancer Stages, Lectures on Modeling and simulation, Trivandrum, India December 1-3, 2009 Vol. 10 Issue1 2009 Published AMSE Journal: 2009 Vol. 10 - N° 1 pp 94-101.

[2] Hermanek P, Sobin "International union Against Cancer TNM classification of malignant tumors, 4th Ed, 2nd revision" LH Editors, Berlin, Springer- Verlag; 1992.

[3] American Cancer Society, Cancer Facts and figures, Atlanta (GA), the society, 1996.

[4] D. M. Parkin, P. Pisani, J. Ferlay, Estimates of the worldwide incidence of twenty five major cancers in 1990. Int J Cancer, Vol.80, 1999, pp827–41.

[5] S. R. Aziz, Oral submucous fibrosis: an unusual disease, J N J Dent Assoc, Vol. 68, 1997, 17–19.

[6] R. B. Zain, N. Ikeda, I. A. Razak, A national epidemiological survey of oral mucosal lesions in Malaysia, Community Dent Oral Epidemiol, Vol. 25, 1997, pp.377–83.

[7] M. D. Rosmai, A. K. Sameemii, A. Basir, I. S. Mazlipahiv, M. D. Norzaidi, The Use of Artificial Intelligence to Identify People at Risk of Oral Cancer: Empirical Evidence in Malaysian University, International Journal of Scientific Research in Education, Vol.3, No.1,

[8] 2010, pp.10-20

[9] D.S.V.G.K. Kaladhar, B. Chandana, P.B. Kumar, Predicting cancer survivability using Classification algorithms, International Journal of Research and Reviews in Computer Science, Vol.02, No.02, 2011, pp.340–343.

[10] M. D. Rosmai, A. Basir, S. A. Kareem, S. M. Ismail, M. D. Norzaidi, Determining the Critical Success Factors of Oral Cancer Susceptibility Prediction in Malaysia Using Fuzzy Models, Sains Malaysiana, Vol.41, No.5, 2012, pp.633–640.

[11] J. O. Kang, S. H. Chung, Y. M. Suh, Prediction of Hospital Charges for Cancer Patients with Data Mining Technique, J Kor Soc Med Informatics, Vol.15, No.1, 2009, pp.13-23.

[12] N. Sharma, H. Om, Framework for early detection and prevention of oral cancer using data mining, International Journal of Advances in Engineering and Technology, Vol.4, No.2, 2012, pp.302-310.

[13] Woonggyu Jung, Jun Zhang, Jungrae Chung, Petra Wilder – Smith, Matt Brenner, J. Stuart Nelson and Zhongping Chen, (2005) "Advances in Oral Cancer Detection using Optical Coherence Tomography", IEEE Journal of Selected Topics in Quantum Electronics, Vol. 11, No.4.pp 811 – 817.

[14] Simon Kent, "Diagnosis of oral cancer using Genetic Programming – A Technical Report", CSTR -96-14.

[15] A. Chodorowski, U. Mattsson, T. Gustavsson, "Oral Lesion classification using true color images", Proceedings of SPIE, Vol. 3661, ISBN. 978081943132, pp 1127 – 1138, 1999.

[16] M. Muthu Rama Krishnan, Chandran Chakraborthy, Ajoy Kumar Ray, "Wavelet based texture classification of oral histopathological sections", International Journal of Microscopy, Science, Technology, Applications and Education, pp 897-906.

[17] Neha Sharma, Nigdi Pradhikaran, Akurdi, "Comparing the performance of data mining techniques for oral cancer prediction", Proceedings of the 2011 International Conference on Communication, Computing & Security (ICCCS'11), ISBN: 978-1-4503-0464-1, New York, USA, 2011.
[18] Yung –nien Sun, Yi-ying Wang, Shao-chien Chang, Li-wha Wu and Sen – tien Tsai, "Color – based tumor segmentation for the automated estimation of oral cancer parameters", Microscopy Research and Technique, Vol. 73, Issue. 1, pp 5- 13, 2010.
[19] Ranjan Rashmi Paul, Anirban Mukherjee, Pranab K. Dutta, Swapna Banerjee, Mousumi, Pal, Jyotirmoy Chatterjee and Keya Chaudhuri, "A novel wavelet neural network based pathological stage detection technique for an oral precancerous condition" , Journal of Clinical Pathology, Vol.58, Issue.9, pp 932 – 938, 2005.
[20] Aeinfar V., Mazdarani H., Deregeh F., Hayati M., and Payandeh M., "Multilayer Perceptron Neural Network with Supervised Training Method for Diagnosis and Predicting Blood Disorder and Cancer," *in Proceedings of IEEE International Symposium on Industrial Electronics*, Korea, pp.2075-2080, 2009.
[21] Sunil Kumar Prabhakar, Harikumar Rajaguru, "GMM Better than SRC for Classifying Epilepsy Risk Levels from EEG Signals", Proceedings of the International Conference on Intelligent Informatics and BioMedical Sciences (ICIIBMS), November 28-30, Okinawa, Japan.
[22] C.-K. Siew and G.-B. Huang. (2005). Extreme Learning Machine with Randomly Assigned RBF Kernels, *Int'l J. Information Technology,* vol. 11, no. 1, 2005.

Measurement of Pulse Transit Time (PTT) Using Photoplethysmography

Lokharan M, Lokesh Kumar K C, Harish Kumar V, Kayalvizhi N *, Aryalekshmi R *

Department of Electronics and Communication Engineering,
Amrita School of Engineering, Coimbatore,
Amrita Vishwa Vidyapeetham,
Amrita University, India
n_kayalvizhi@cb.amrita.edu, r_aryalekshmi@cb.amrita.edu

Abstract. An experimental setup for measuring pulse transit time (PTT) using two photoplethysmographic (PPG) signals is proposed. The pulse wave velocity is calculated from the measured PTT, for its potential use in monitoring arterial stiffness and continuous non-invasive arterial blood pressure. The PPG sensors are placed in-line separated by 1cm, over the arterial branch under consideration. PTT is determined by using the fact that there exists a time difference be-tween PPG signals obtained from the two sensors. The outputs from the sensors are bandlimited and amplified. The signals are then made TTL- compatible, to determine pulse transit time. The final output is in the form of a rectangular pulse whose width is equal to the PTT. A phase shifted sinusoid generated by an all-pass filter is used to validate the proposed experimental setup. Placing the sen-sors over the digital artery of the index finger, 1 cm apart, the PTT for 13 healthy volunteers was measured. PTT is averaged for 10 cardiac cycles to obtain mean PTT. From mean PTT the Pulse Wave Velocity (PWV) is calculated for a volunteer. The maximum of 2.6 and a minimum of 0.4 standard deviation was observed, for a single volunteer. The overall mean PTT was 10.1 ms and the mean PWV was 1.1 m/s.

Keywords: Pulse Transit time, Wave reflections; Inline sensors; Aurdino

1 Introduction

Photoplethysmography is a widely accepted clinical method to measure the physiological parameters such as cardiac output, oxygen saturation. Its compactness and cost effectiveness makes it a better option for studying cardio vascular health [2]. Also studies prove that PPG can be obtained calibration free [8]. PPG can be used to determine an important parameter called Pulse Wave Velocity (PWV). PWV can be used to infer information about arterial stiffness and blood pressure. Algorithms have been proposed to estimate arterial blood pressure from the PWV [1]. PWV can be determined with the help of Doppler ultrasound technique [11]. But the apparatus for Doppler ultrasound is expensive. Two other methods have also been proposed for measuring PWV from PPG. Pulse transit time is considered to be the time between peak in PPG signal and R peak in ECG signal [3]. The drawback of this method is that there is a finite time delay between the actual ventricular compression and R- peak in ECG [4]. PTT can be measured, by measuring the time difference between the two PPG signals from the in-line sensors placed at a distance apart [5]. The obtained time difference is Pulse Transit Time. The PWV can be calculated using (1).The inline sensors are positioned such that, one sensor is placed at wrist and the other at the foot of pinky finger. The PPG sensor at the wrist senses PPG in radial artery, and sensor in the pinky finger senses the signal from digital artery. Near the wrist the radial artery divides into five digital arteries. In this method, measuring the PWV over long arterial lengths can lead to ambiguity over the path lengths. Further there may be influence of blood wave reflections due to branching of the artery [6]. To address the above issues, the PWV is measured over small arterial lengths where the sensors are placed at small distance apart. This PWV is known as the local PWV [7].In this paper, design of a hardware module to measure the PTT and hence the local PWV over the digital artery in the index finger is proposed. The circuit is automated to measure PTT and hence can be a potential tool for monitoring arterial stiffness and continuous non-invasive measurement of blood pressure.

2 Determination of PTT

The block diagram for the hardware module is shown in the Fig. 1. It is to be ensured that the response of the two circuits must be identical i.e. the phase shift induced on a signal by circuit 1 must be equal to the phase shift induced on the same signal by circuit 2. This is necessary because the circuit giving non-identical phase shifts will impart additional phase which will be a serious source of error.

J. Goh et al. (eds.), *The 16th International Conference on Biomedical Engineering*,
IFMBE Proceedings 61,
DOI: 10.1007/978-981-10-4220-1_24

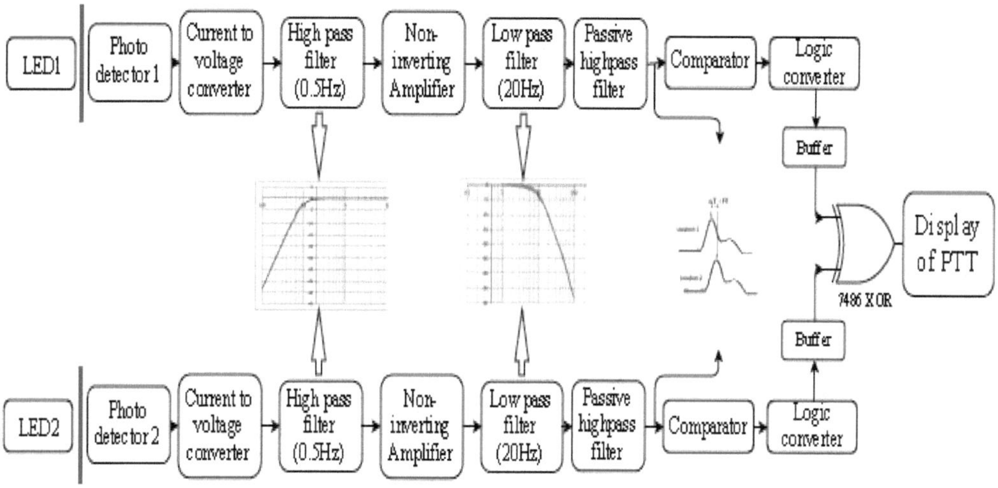

Fig. 1: The circuit block diagram

Performance of the hardware is verified in two stages. The identical working of both PPG circuits is verified by analysing their frequency response plots. The correctness of the PTT calculation is verified by an all-pass filter. The all-pass filter generates the phase shifted version of the input sinusoid. Both the original sinusoid and the phase shifted version of it is given to the circuit module and a time difference is calculated. This time difference must correspond to the phase shift induced the all-pass filter. The methodology for obtaining the PPT is explained in the following sections. The circuit employs two identical circuits to acquire PPG from two sensors placed in-line on an arterial site. The circuit involves an LED excitation circuit .

2.1 Acquiring PPG

The whole design is based on reflectance type PPG. The PPG signals are acquired from the digital artery using two sensors. The sensor constitutes an IR LED and a photodiode adjacent to each other. The sensor used is shown in Fig. 2. The IR led is used because light has maximum penetration in human tissue for the wavelength in the range of 650 to 1350 nm [9]. Also the photodiode used has maximum response for IR LED. The voltage across the LED is regulated using a zener diode. The current through the LED can be varied by a potentiometer in series with the LED. The blood in arteries reflects back the IR light in accordance with the pulsatile flow within. The reflected light falls on the photo-

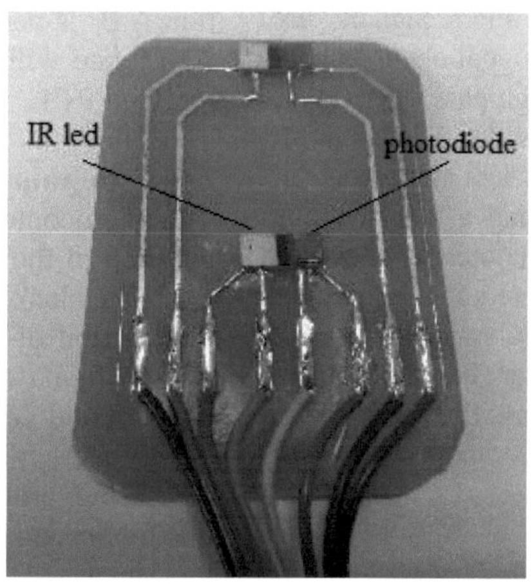

Fig. 2: DCM03 sensors.

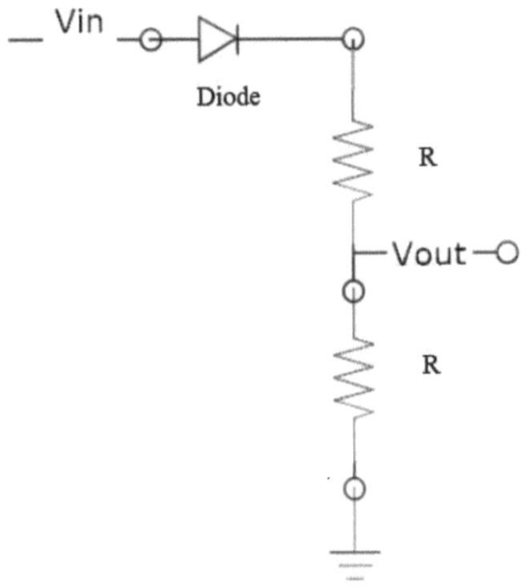

Fig. 3: Level conversion circuit used.

diode and generates a current which corresponds to the intensity of the reflected light.

2.2 Current to voltage conversion

The photodiode is connected to an I to V converter which is realized using an op-amp with a RC feedback network to give a voltage proportional to the input current from the photodiode. The transresistance of the amplifier is 220kΩ. The photo-diode is biased in photo-voltaic mode.

2.3 Bandlimiting the signal

The PPG signal obtained from I to V converter has contributions from many frequency components, but dominant contribution is from the components with frequency less than 20Hz. Also, the DC voltage has to be removed to identify the zero crossing points correctly. Hence the PPG signals are bandlimited to 0.5 to 20 Hz. The voltage signal obtained contains noise and is first passed through a high pass filter (HPF) to remove DC voltage. The HPF is designed for 0.5 Hz cut-off frequency. Since the voltage levels are very less at the range of few millivolts, the signal is amplified using a non-inverting amplifier. The amplifier is designed using a potentiometer with which the gain can be varied according to required amplitude. Further it is passed through a low pass filter having a cut-off at 20Hz. Both the filters are second order filters based on sallen-key topology [10]. Another passive high pass filter is added to further attenuate the very low frequency components and the effects of DC induced by the op-amps.

2.4 Conversion to square wave

Upon observing on an oscilloscope, a minute phase difference should be observed between the signals which is to be determined. To find the PTT, the signals are passed to a comparator to convert them into square pulses. Upon conversion, we obtain two square pulses with a small phase difference. These signals have voltage levels of +10V and -10V. Since we are using digital logic gates, the signals have to be made compatible. So, the signals are converted to the compatible logic of +5V and 0V using the circuit shown in Fig. 3.

2.5 Determining the PTT

The signal from the lagging sensor is negated using a digital NOT logic gate. Further, digital AND logic is performed i.e. considering the PPG waveforms as A and B, digital operation A.B' is performed using logic gates. This output of the AND gate will be a rectangular pulse whose width is the required PTT.

2.6 Programming the Microcontroller

The microcontroller, Arduino UNO, is used to measure the duration of the pulse obtained from the digital logic operation. The clock frequency of the UNO board is 16MHz. The time period of its clock is 0.0625us.The Pulse transit time can be found as a product of the counter output and the clock period i.e. 0.0625 us. A small delay of 300ms is given between the measurements of successive samples to avoid measuring unwanted transitions which occurs sometimes due to dicrotic notch of PPG signal. The UNO board is further interfaced with a 16x2 LCD display to display the measured time.

2.7 Calculation of Pulse Wave Velocity

The PWV was calculated by the formula

$$PWV = D / PTT \quad (1)$$

Where D is the distance between the sensors. The sensors are placed at a distance of 1cm.

3 Experimental validation

3.1 In-vivo measurements

The PTT was measured for 13 healthy volunteers in the age group of 21-35 years. Also their PWV was calculated

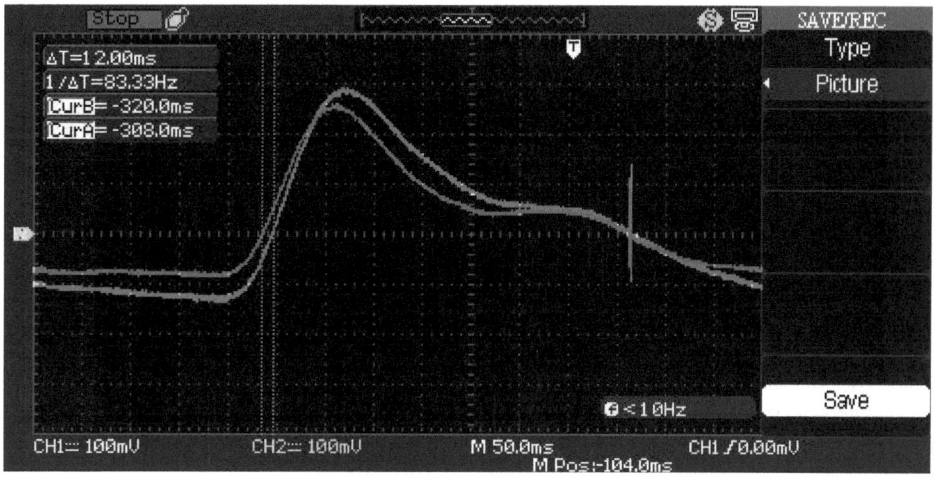

Fig. 4: PPG waveforms of volunteer A

Table 1. Calculated Mean PTT, standard deviation and PWV

Subject	Mean PTT (ms)	Std. deviation	PWV (m/s)
1	8.3	0.9	1.2
2	6.7	0.7	1.5
3	6.9	0.5	1.4
4	7.2	0.4	1.4
5	8.1	1.2	1.2
6	8.5	0.8	1.2
7	9.3	0.7	1.1
8	9.5	2.6	1.1
9	10.2	1.7	1.0
10	11.4	0.9	0.9
11	11.6	0.8	0.9
12	11.6	0.7	0.9
13	12.1	0.9	0.8

using (1). Mean and standard deviation were calculated for 10 PTT samples for every volunteer. The signals were acquired from the radial artery of the index finger. The Fig.4 shows the PPG waveform of volunteer A and the cursor measuring the PTT. The calculated results and measured results have been tabulated.

3.2 Validation using phase shifted sinusoid

The circuits were given phase shifted sinusoids as inputs. The required phase difference was achieved using an all-pass filter. The measured time difference was compared with the phase difference induced by the all-pass filter and has a minimum error of 0.03ms to a maximum error of 0.9ms.

4 Conclusions

The design of a hardware module for measurement of Pulse Transit Time from two PPG signals was presented. A maximum and minimum standard deviation of 2.6 and 0.4 respectively was observed, for single volunteer. The mean PTT for 13 volunteers was determined to be 10.1 ms and the calculated mean PWV was 1.1m/s. The validation of circuit was done by using the phase shifted sinusoids and the output of the circuit matched the phase difference induced by the all-pass filter. The circuit could respond to small variations in PTT thus indicating its potential use in monitoring of arterial stiffness and blood pressure non-invasively on a continuous basis.

References

[1] D. B. McCombie, A. T. Reisner and H. H. Asada, "Adaptive blood pressure estimation from wearable PPG sensors using peripheral artery pulse wave velocity measurements and multi-channel blind identification of local arterial dynamics," *Engineering in Medicine and Biology Society, 2006. EMBS '06. 28th Annual International Conference of the IEEE*, New York, NY, 2006, pp. 3521-3524.

[2] J. Allen, "Photoplethysmography and its application in clinical physiological measure-ment", *Physiol. Meas.*, vol. 28, no. 3, pp. R1-R39, 2007.

[3] R. Payne, "Pulse transit time measured from the ECG: an unreliable marker of beat-to-beat blood pressure", *Journal of Applied Physiology*, vol. 100, no. 1, pp. 136-141, 2006.

[4] P. Fung, G. Dumont, C. Ries, C. Mott, M. Ansermino, "Continuous Noninvasive Blood Pressure Measurement by Pulse Transit Time," Proceedings of the 26th Annual International Conference of the IEEE EMBS San Francisco, CA, USA € September 1-5, 2004.

[5] K. Hirata, M. Kawakami and M. O'Rourke, "Pulse Wave Analysis and Pulse Wave Veloci-ty", *Circulation Journal*, vol. 70, no. 10, pp. 1231-1239, 2006.

[6] P. Segers, J. Kips, B. Trachet, A. Swillens, S. Vermeersch, D. Mahieu, E. Rietzschel, M. De Buyzere and L. Van Bortel, "Limitations and pitfalls of non-invasive measurement of arterial pressure wave reflections and pulse wave velocity", *Artery Research*, vol. 3, no. 2, pp. 79-88, 2009.

[7] P. M. Nabeel, J. Joseph and M. Sivaprakasam, "Magnetic plethysmograph transducers for local blood pulse wave velocity measurement," *2014 36th Annual International Conference of the IEEE Engineering in Medicine and Biology Society*, Chicago, IL, 2014, pp. 1953-1956.

[8] H. Harini, L. Krithika, M. Shalini, S. Swaminathan and N. Madhu Mohan, "Design and Implementation of a Calibration - Free Pulse Oximeter", *The 15th International Conference on Biomedical Engineering*, Singapore, 2013, pp. pp 100-103.

[9] Smith, Andrew M.; Mancini, Michael C.; Nie, Shuming (2009). "Bioimaging: Second window for in vivo imaging". Nature Nanotechnology **4** (11): 710–711.

[10] Sergio Franco," Design with operational amplifiers and analog integrated circuits," 3rd edition, Tata McGraw- Hill Edition, 2002, pp. 106-141.

[11] Beulen, Bart WAMM, et al. "Toward noninvasive blood pressure assessment in arteries by using ultrasound." *Ultrasound in medicine & biology* 37.5 (2011): 788-797

Erratum to: The 16th International Conference on Biomedical Engineering

James Goh, Chwee Teck Lim, and Hwa Liang Leo

Department of Biomedical Engineering,
National University of Singapore,
Singapore

Erratum to:
J. Goh et al. (eds.),
The 16th International Conference on Biomedical Engineering,
IFMBE Proceedings 61,
DOI: 10.1007/978-981-10-4220-1

The original version of the book was inadvertently published with incorrect title "The 17th International Conference on Biomedical Engineering" which has to be corrected to read as "The 16th International Conference on Biomedical Engineering". The erratum book has been updated with the change.

The updated online version of this book can be found at http://dx.doi.org/10.1007/978-981-10-4220-1

J. Goh et al. (eds.), *The 16th International Conference on Biomedical Engineering*,
IFMBE Proceedings 61,
DOI: 10.1007/978-981-10-4220-1_25

Author Index

A
Abhay, Theertha 47
Akulov, S.A. 16, 95
Akulova, A.S. 16, 95
Ali, M.A.M. 32
Asako, Mikiya 53

B
Bharathi, Kunal 32
Born, Jannis 43

C
CH, Nikhil 88
Che, U. Kin 111

D
Delrieu, A. 58
Doyle, D. John 28

F
Fedotov, A.A. 16, 95

G
Gaonkar, Manjunath 32

H
Hassani, Kamran 28
Hoël, M. 58

I
Ichikawa, Tomu 1
Inou, Norio 1
Inoue, Wataru 1

J
J., Rolant Gini 47, 106
Jafarian, Kamal 28
Jiang, Xiao 111

K
K.C., Lokesh Kumar 130
K.I., Ramachandran 99, 106
Karun, Sai 32
Kimura, Hitoshi 1
Kirthika, S.K. 117
Kobayashi, Yoshiki 53
Krivokuca, Sacha 1
Krupa, B. Niranjana 32
Kuramoto, Akisue 1

L
Lissorgues, G. 58
Liu, Xin 111

M
M., Lokharan 130
Mak, Peng Un 111
Malini, M. 88
Miyamoto, Yuichi 13

N
N., Kayalvizhi 47, 130
Nair, Aishwarya 117
Nath, Suhan 32

O
Ono, Hiroyuki 1
Otake, Yasuhito 1

P
Phua, C.T. 58
Prabhakar, Sunil Kumar 21, 73, 80, 123
Pun, Sio Hang 111

R
R., Aishwarya 99
R., Aryalekshmi 130
R., Lavanya 62
Rajaguru, Harikumar 21, 73, 80, 123
Ramachandran, Babu Ram Naidu 43
Ratnam, Rama 43
Reshetnikova, M.A. 16, 95
Romero Pinto, Sandra Alejandra 43

S
Sekiyama, Naoto 1
Shashikumar, Niranjana 117
Shikano, Shunpei 53
Simon, Sneha 62
Suman, Dabbu 88
Sun, Guanghao 38
Suzuta, Haruna 7

T
T., Vasundhara 99
Tajima, Hatsuki 13
Takeyama, Masahiro 53
Tanaka, Toshiyuki 7
Tomoda, Koichi 53

U
U.K., Ceerthibala 106

V
V., Harish Kumar 130
Vai, Mang I. 111
Vamsi, Reddy 88
Vidhya, S. 117
Vijayan, Devi 62
Vo, Tri Thong 67

W
Wang, Jiujiang 111
Winkler, Stefan 43

Y
Yamaguchi, Yuri 13
Yamamoto, Takahisa 53
Yu, Wenwei 38
Yu, Yuanyu 111

MIX
Papier aus verantwortungsvollen Quellen
Paper from responsible sources
FSC® C105338

If you have any concerns about our products,
you can contact us on
ProductSafety@springernature.com

In case Publisher is established outside the EU,
the EU authorized representative is:
Springer Nature Customer Service Center GmbH
Europaplatz 3, 69115 Heidelberg, Germany

Printed by Libri Plureos GmbH
in Hamburg, Germany